# Managing Intellectual Property Relevant to Operating and Sustaining Major U.S. Air Force Weapon Systems

FRANK CAMM, PHILLIP CARTER, SHENG TAO LI, MELISSA SHOSTAK

Prepared for the Department of the Air Force
Approved for public release; distribution unlimited

RAND PROJECT AIR FORCE

For more information on this publication, visit **www.rand.org/t/RR4252**.

**About RAND**

The RAND Corporation is a research organization that develops solutions to public policy challenges to help make communities throughout the world safer and more secure, healthier and more prosperous. RAND is nonprofit, nonpartisan, and committed to the public interest. To learn more about RAND, visit www.rand.org.

**Research Integrity**

Our mission to help improve policy and decisionmaking through research and analysis is enabled through our core values of quality and objectivity and our unwavering commitment to the highest level of integrity and ethical behavior. To help ensure our research and analysis are rigorous, objective, and nonpartisan, we subject our research publications to a robust and exacting quality-assurance process; avoid both the appearance and reality of financial and other conflicts of interest through staff training, project screening, and a policy of mandatory disclosure; and pursue transparency in our research engagements through our commitment to the open publication of our research findings and recommendations, disclosure of the source of funding of published research, and policies to ensure intellectual independence. For more information, visit www.rand.org/about/principles.

RAND's publications do not necessarily reflect the opinions of its research clients and sponsors.

Published by the RAND Corporation, Santa Monica, Calif.
© 2021 RAND Corporation
**RAND®** is a registered trademark.

Library of Congress Cataloging-in-Publication Data is available for this publication.

ISBN: 978-1-9774-0780-1

*Cover: Senior Airman Peter Reft.*

# Preface

In 2018, RAND Project AIR FORCE (PAF) delivered a report to the Air Force Special Operations Command (AFSOC) that detailed issues that AFSOC currently faces regarding technical data and data rights.[1] At the request of the Office of Air Force Deputy General Counsel for Acquisition, PAF extended this research to address the findings of studies mandated by Sections 809, 813, and 875 of the Fiscal Year 2016 National Defense Authorization Act, which were released from 2017 through 2019 and related to the subject of the AFSOC-sponsored research.

This report describes ways for the Air Force to improve its policy on and management of technical data and data rights relevant to operating and sustaining major weapon systems.[2] Many of the problems addressed and recommendations developed may be applicable to other types of technical data and data rights problems throughout the Department of Defense. It should interest policymakers and analysts in these organizations responsible for work on technical data and data rights policy. By design, this document does not cover other intellectual property issues of interest to the Air Force. For example, it does not address trademarks registered by the Air Force or patents arising from organic Air Force–funded research.

The research reported here was commissioned by the Office of the Air Force Deputy General Counsel for Acquisition and conducted within the Resource Management Program of PAF as part of a fiscal year 2019 project, Analysis of Air Force Acquisition of Intellectual Property Rights.

## RAND Project AIR FORCE

PAF, a division of the RAND Corporation, is the U.S. Air Force's federally funded research and development center for studies and analyses. PAF provides the Air Force with independent analyses of policy alternatives affecting the development, employment, combat readiness, and support of current and future air, space, and cyber forces. Research is conducted in four programs: Strategy and Doctrine; Force Modernization and Employment; Manpower, Personnel, and Training; and Resource Management. The research reported here was prepared under contract FA7014-16-D-1000.

---

[1] Frank Camm et al., *Data Rights Relevant to Weapon Systems in Air Force Special Operations Command*, Santa Monica, Calif.: RAND Corporation, RR-4298-AF, 2021.

[2] Note that this report was written before the U.S. Space Force was created within the Department of the Air Force.

Additional information about PAF is available on our website:
www.rand.org/paf/

This report documents work originally shared with the U.S. Air Force on March 26, 2019. The draft report, issued on October 25, 2019, was reviewed by formal peer reviewers and U.S. Air Force subject-matter experts.

# Contents

# Figure

# Tables and Boxes

# Summary

## Issue

The Air Force has found that, without appropriate rights and access to intellectual property (IP), the service faces constraints in sustaining, modifying, and modernizing its systems. The IP problems differ for new versus legacy systems; new IP strategies can be crafted for new systems, but legacy systems must live with IP decisions made early in those systems' life cycles. The question presented for this report is how the Air Force can best acquire, use, and leverage IP for new and legacy systems, operating within current legal and contractual constraints, and what organizational change the Air Force should consider to better acquire and use IP.

## Approach

PAF reviewed recently completed studies of this problem and drew on recent RAND analysis to identify issues and potential solutions on which senior Air Force leaders should place the greatest emphasis. To help identify these issues, PAF analyzed recent congressionally directed work and worked closely with ongoing efforts in the Air Force legal, acquisition, and sustainment communities to improve its IP policy and practices.[1]

## Conclusions

- There is general agreement that Air Force efforts to expand its reliance on original equipment manufacturers (OEMs) to sustain its weapon systems in the 1980s and 1990s did not work as well as expected. Sustainment by OEMs has typically been more costly, less responsive, and of lower quality than the Air Force believes in-house or competitively provided sustainment would have been.
- In-house and competitive sustainment of Air Force systems requires Air Force access to appropriate IP. Expanded reliance on OEMs led the Air Force to give less priority to acquiring appropriate IP. This change in priorities reduced the skills within the Air Force

---

[1] Advisory Panel on Streamlining and Codifying Acquisition Regulations, *Interim Report*, Washington, D.C.: U.S. Department of Defense, May 2017; *Volume 1*, January 2018; *Volume 2*, June 2018; *Volume 3*, January 2019 (Section 809 Panel; hereafter, Section 809 Panel Report); U.S. Department of Defense, *2018 Report of the Government-Industry Advisory Panel on Technical Data Rights*, Washington, D.C.: U.S. Department of Defense, November 13, 2018 (hereafter, Section 813 Panel Report); Richard Van Atta et al., *DoD Access to Intellectual Property for Weapon Systems Sustainment*, IDA Paper P-8266, Alexandria, Va.: Institute for Defense Analyses, May 2017.

to identify requirements for IP and to translate these requirements into effective delivery and then management of technical data and data rights.

- Air Force IP specialists understand these issues in some depth, but the Air Force (like other services) currently has few of these specialists. Congress directed the establishment of an IP cadre at the Office of the Secretary of Defense or service level. Several models of cadre suggest that an IP cadre should have specialized knowledge, be centralized, have a focused mission, be funded by programs, and be supported by Air Force leadership.
- Senior Air Force leaders will need to actively promote any new initiatives to acquire appropriate IP for new systems or to work around the general dearth of appropriate IP for legacy systems. They will need to sustain a multiyear effort to staff, resource, and coordinate new policies and practices across the Air Force legal, acquisition, and sustainment communities to assure more cost-effective support to the warfighter.

## Recommendations

- Structure qualitatively different approaches to IP policy and practice for legacy and new Air Force systems.
- Understand the role of time when comparing the cost and benefits of investing in IP. It is more cost-effective for the Air Force to invest in IP if the real discount rate that it uses to make decisions, adjusted for inflation, is lower than that of the relevant OEM.
- Clarify the scope of operations, maintenance, installation, and training and form-fit-function data in service and/or Department of Defense regulations and contract clauses.
- Develop a cadre of IP specialists, and motivate its effective use throughout the Air Force.
- Create a standard mechanism (such as an advance agreement on pricing) to preserve an option to acquire IP at a price negotiated as part of the engineering and manufacturing development (EMD) source selection.
- If Air Force leadership decides it is worthwhile to pursue IP over the long term, commit to doing this in a coordinated manner in each phase of the acquisition of a program.
- When appropriate, motivate relevant Air Force personnel to commit the resources they control to more proactive pursuit of effective Air Force access to IP in major programs.
- Use formal change management to coordinate the diverse activities necessary to improve IP policy.
- Continue research and policy development relating to IP strategy, acquisition, use, and management by the Air Force.

**Box S.1. Illustrative Opportunities for Use of Intellectual Property**

| Legacy Systems |
| --- |

- Procurement of additional IP to facilitate organic support or depot-level maintenance
- Procurement of additional IP to facilitate contracting with third-party contractor for support or depot-level maintenance
- Procurement of additional IP to facilitate upgrades, modifications, and service-life extension for systems and subsystems
- Forensic auditing of legacy contracts to determine existing rights in IP
- Procurement of additional platforms

| New Systems |
| --- |

- Choice of license for Air Force rights in technical data and rights in software
- Creation of contract deliverables to ensure delivery of IP
- Engagement in markings disputes to clarify Air Force IP rights during EMD contract performance
- Development of Air Force doctrine, operational concepts, training manuals, etc.
- Inclusion of new systems' intellectual data in Air Force business information systems (i.e., predictive analytics of maintenance and supply needs)

# Acknowledgments

Joseph M. McDade, Jr., Principal Deputy General Counsel of the Air Force, sponsored this work. Mark Borowski, Nancy Kremers, and David Ruddy, in the Office of the Air Force Deputy General Counsel for Acquisition; Lawrence S. Kingsley, Deputy Assistant Secretary of the Air Force for Logistics and Product Support; and Michael Oar, in Lawrence S. Kingsley's office, provided valuable insights and support throughout the project. The personnel at Hill Air Force Base provided us with access and additional understanding of the problems addressed within the report.

At RAND, Obaid Younossi oversaw this project and provided sage advice and support throughout. Geoff McGovern and Megan McKernan provided valuable commentary on a prior study, which greatly informed the content and direction of this project.

We thank everyone for their help. Any factual errors or misjudgments remain our responsibility.

# Abbreviations

AFSOC          Air Force Special Operations Command

CAO          Competency Aligned Organization

CDRL          contract data requirements list

CLIN          contract line item number

DCAA          Defense Contract Audit Agency

DCMA          Defense Contract Management Agency

DFARS          Defense Federal Acquisition Regulation Supplement

DoD          Department of Defense

EMD          engineering and manufacturing development

FAA          Federal Aviation Administration

FAR          Federal Acquisition Regulation

FFF          form-fit-function

FFRDC          federally funded research and development center

IDA          Institute for Defense Analysis

IG          Inspector General

IP          intellectual property

IPR CFT          Intellectual Property Rights Cross Functional Team

IPT          integrated product or project team

ISO          International Organization for Standardization

JCIDS          Joint Capabilities Integration and Development System

KPP          key performance parameter

NAVAIR          Naval Air Command

NDAA          National Defense Authorization Act

NPV          net present value

O&S          operations and support

OEM          original equipment manufacturer

| | |
|---|---|
| OMB | Office of Management and Budget |
| OMIT | operation, maintenance, installation, or training |
| OSD | Office of the Secretary of Defense |
| PAF | Project AIR FORCE |
| R&D | research and development |
| RFP | request for proposal |
| SAF/AQ | Assistant Secretary of the Air Force for Acquisition |
| SAF/GC | Air Force General Counsel |
| SNLR | specially negotiated license right |

# 1. Introduction

In recent years, interest in improving access to technical data and data rights associated with major Department of Defense (DoD) programs has suddenly grown. This burst in interest appears to be the product of a growing recognition throughout DoD that the department does not have access to the technical data and data rights that it needs to implement changes in policy that could potentially improve (and reduce costs for) operations and support (O&S); execution of upgrades, modernization, and life extension; and long-term sustainment of the supply chain for its major programs. This report is a response to the Air Force's increasing concerns about its access to such technical data and data rights.[1]

In February 2018, Under Secretary of the Air Force Matthew P. Donovan established the Intellectual Property Rights Cross-Functional Team (IPR CFT) to address a broad set of questions about how the Air Force could improve its acquisition and use of intellectual property (IP), particularly that associated with technical data and data rights relevant to major Air Force programs. He gave the Assistant Secretary of the Air Force for Acquisition (SAF/AQ) and the Air Force General Counsel (SAF/GC) responsibility to lead this team.[2]

This activity joined others initiated by the National Defense Authorization Act (NDAA) for Fiscal Year 2016 (Public Law 114–92) to address similar issues. Section 809 mandated the creation of an independent Advisory Panel on Streamlining and Codifying Acquisition Regulations. This came to be known as the Section 809 Panel. It produced a series of reports in May 2017, January 2018, June 2018, and January 2019.[3] Section 813 directed the Secretary of Defense to establish a government-industry advisory panel on technical data rights. This came to be known as the Section 813 Panel. It produced a final report in November 2018.[4] And Section 875 of the FY 2016 NDAA mandated a federally funded research and development center (FFRDC) study of "Department of Defense (DOD) regulations, practices, and sustainment

---

[1] This report focuses on intellectual property for major weapon systems, based on the scope of the project description agreed on by the Air Force and RAND. This report does not address the many other types or uses of intellectual property relating to the Air Force, such as innovations in doctrine or operational concepts, intellectual property relating to commercial systems (such as commercial software or commercial off-the-shelf hardware), or trademark and licensing issues that arise during acquisition. Note that this report was written before the U.S. Space Force was created within the Department of the Air Force.

[2] IPR CFT had representatives from SAF/AQ, SAF/GC, Headquarters Air Force Judge Advocate Group. Air Force Space Command and Air Force Materiel Command. Matthew P. Donovan, "Intellectual Property Rights Cross-Functional Team (IPR CFT)," memorandum, Washington, D.C.: Office of the Under Secretary of the Air Force, February 21, 2018.

[3] Section 809 Panel Report.

[4] Department of Defense, *2018 Report of the Government-Industry Advisory Panel on Technical Data Rights* (hereafter, Section 813 Panel Report), Washington, D.C.: Department of Defense, November 13, 2018.

requirements related to government access to and use of IP rights of private-sector firms and . . . DOD practices related to the procurement, management, and use of intellectual property rights to facilitate competition in sustainment of weapon systems throughout their lifecycle." The Institute for Defense Analyses (IDA) produced this report in May 2017.[5]

The Office of the Air Force General Counsel approached RAND's Project AIR FORCE (PAF) in July 2018 to help the Air Force synthesize findings and recommendations being developed in these panels and studies and to clarify their implications for the Air Force. PAF worked closely with the personnel with day-to-day responsibility for the operations of the Air Force IPR CFT to design a product that would rise above the technical details produced by the panels and studies described above to identify higher-level issues that warranted the attention of the senior leadership of the Air Force. This report is that product.[6]

## Road Map

This chapter introduces the report and its context. Chapter 2 explains why the Air Force thinks it needs more technical data and data rights. It explains how the challenge of getting IP differs in the legacy programs that currently dominate Air Force budgets and the new programs where the Air Force potentially has better leverage to get better data and rights.

Chapter 3 explains that efforts to get data and rights face a general problem that exists throughout the Air Force—wherever the Air Force must make investments today to get benefits in the future, often the far future. Many factors have led the Air Force to give long-term benefits less attention relative to current costs than may have been appropriate. Until the Air Force leadership can address this broader problem, it will have difficulty getting the technical data and data rights that many leaders believe the Air Force needs today.

Chapter 4 examines the life cycle of a major program to identify factors that make it difficult for the Air Force to get more complete technical data and data rights. Challenges arise throughout the life cycle. This chapter explains that the Air Force must make decisions very early in a program to get the data and rights it can benefit from when it operates and supports a major system decades later. When it does the right things up front, it must follow up and manage its access to data and rights to use them throughout the life cycle. Coordination throughout the life cycle requires the alignment of many Air Force communities to common goals.

Chapter 5 describes the way formal change management has helped complex, technologically sophisticated global organizations align many communities within themselves to change in ways

---

[5] Richard Van Atta et al., *DoD Access to Intellectual Property for Weapon Systems Sustainment*, IDA Paper P-8266, Alexandria, Va.: Institute for Defense Analyses, May 2017.

[6] Other work conducted for the Air Force Special Operations Command helps frame our approach to the analysis reported here. Frank Camm, Thomas C. Whitmore, Guy Weichenberg, Sheng Tao Li, Phillip Carter, Brian Dougherty, Kevin Nalette, Angelena Bohman, and Melissa Shostak, *Data Rights Relevant to Weapon Systems in Air Force Special Operations Command*, Santa Monica, Calif.: RAND Corporation, RR-4298-AF, 2021.

that promote common, enterprisewide goals. It examines such change from the perspective of relevant communities in the Air Force and explains how these communities can address their specific data-related issues in ways that remain aligned to Air Force–wide priorities regarding technical data and data rights.

Chapter 6 addresses a proposal highlighted in two of the congressionally mandated efforts described above—the creation of centers of excellence with experts on technical data and data rights who can assist individual program offices throughout the life of a program. It examines four ways that DoD has created and used such cadres in the past and discusses how each might address Air Force efforts to expand access to technical data and data rights.

Chapter 7 addresses another proposal. It seeks to preserve a program's access to technical data and data rights through the life of a program. It discusses the legal challenge of creating options to acquire data and data rights in association with an engineering and manufacturing development (EMD) program and preserving a program's access to these options through operation and support of the systems. It recommends a way to use IP valuation principles commonly applied outside DoD to price such options.

Chapter 8 summarizes our policy recommendations to the senior leadership of the Air Force.

Appendix A summarizes the findings and recommendations of the Section 809 and Section 813 Panel Reports and the Section 875 Study. It offers brief observations that suggest issues that deserve additional attention.

# 2. Why Does the Air Force Need Intellectual Property?

The U.S. Air Force acquires, operates, sustains, and retires a vast array of increasingly sophisticated major weapon systems in defense of the nation. From cradle to grave, in design, manufacture, operation, sustainment, upgrade, or retirement, the creation and use of IP (broadly defined) plays a vital role for each of these systems. Most of this IP originates in the private sector, which the service relies on to imagine, design, and manufacture all of its major weapon systems. This IP runs through each of the major weapon systems acquired by the Air Force. However, beyond that which exists in the physical systems that are procured, the Air Force must also decide what, if any, IP of the contractor it must purchase, and under what terms, to meet Air Force needs. This section describes those needs in more detail, highlighting potential tensions or friction points that may emerge based on experience with Air Force systems.

## Air Force Uses for Intellectual Property

The Air Force has a number of discrete uses for the IP that accompanies its major weapon systems—many of which are in tension with each other. Some of this IP relates to short-term compliance with government requirements (such as the reports delivered under contract deliverable requirements lists during performance of a contract) or such administrative tasks as proposal information given to the government for source selection. Both the short-term compliance and administrative needs and the long-term needs are important. However, most IP relating to weapon systems relates to their design, manufacture, operation, and sustainment. This IP serves at least six discrete needs for the Air Force over the lifetime of these systems.

### Organic Operation and Support

The Air Force requires IP to operate the sophisticated major weapon systems that it acquires. This includes two types of IP statutorily required to be delivered with "unlimited rights" to the government: technical data that "relates to form, fit, or function" or technical data that "is necessary for operation, maintenance, installation, or training."[1] By statute and regulation, the government obtains a license for the unlimited use of this data—sometimes called OMIT (operation, maintenance, installation, or training) data or FFF (form-fit-function) data—because of the importance for government utilization by government personnel of this IP in daily use and sustainment of the system being acquired. Organic operation includes training, operation, and

---

[1] 10 U.S.C. 2320.

4

deployment in operations by government personnel, including, in this case, Air Force military personnel and civilians.

## Third-Party Support

Alongside or in addition to use by Air Force civilian and military personnel, the Air Force may also require third parties (including contractors or allied forces) to operate, maintain, or utilize its major weapon systems. This may include but not be limited to engagement of third-party contractors (other than the original equipment manufacturer [OEM]) to provide maintenance support or employment of contractor support in Air Force depot facilities alongside government civilian and military personnel. Consequently, the Air Force may have a requirement for IP relating to these systems and the legal ability to share that IP with a nongovernment entity. Under the current rights regime relating to technical data on major weapon systems, the Air Force may generally share the OMIT and FFF data that it receives, because it takes an unlimited rights license in that data. It may also, under certain circumstances, share other IP that it obtains under a "government purpose rights" license.[2] However, the default positions in procurement regulations limit the ability of the Air Force to share contractors' IP with third parties, even for official purposes, if it obtains anything less than unlimited rights in that IP.

## Diagnostics and Investigation

In addition to daily operations and maintenance of its aircraft and other major weapon systems, the Air Force requires IP regarding these systems to investigate and diagnose accidents and mishaps in a timely manner, as well as to identify, analyze, and understand issues and trends that may emerge across its fleets of systems. This may include access to design, manufacture, and systems data or software that goes far beyond what is necessary for daily operations and maintenance (and therefore encompassed by OMIT or FFF data). Such a need may also involve the sharing of these data with U.S. government and nongovernment parties, such as the Federal Aviation Administration (FAA) or third-party contractors engaged to help the Air Force with accident investigations, safety and mishap analyses, or analysis of long-term O&S issues.[3] Current procurement law limits the Air Force in its ability to use contractor IP to diagnose and investigate issues within the fleet of aircraft it has procured, except to the extent that it has both acquired sufficient rights in that data and obtained delivery of that data. A large amount of these

---

[2] See DFARS 252.227-7013, described in more detail in U.S. Air Force, Space and Missile Systems Center, *Acquiring and Enforcing the Government's Rights in Technical Data and Computer Software Under Department of Defense Contracts: A Practical Handbook for Acquisition Professionals*, 9th ed., Los Angeles Air Force Base, Calif.: U.S. Air Force, October 2018.

[3] Thomas Light, Dwayne M. Butler, Michael Boito, Vikram Kilambi, Kristin J. Leuschner, Sheng Tao Li, Abby Schendt, and Sunny D. Bhatt, *Management of U.S. Air Force Aircraft Contractor Logistics Support Arrangements: Summary of Findings and Recommendations*, Santa Monica, Calif.: RAND Corporation, RR-4194-AF, forthcoming.

data, including design and manufacture data, as well as operations and maintenance data within the Autonomic Logistics Information System used to instrument and manage certain aircraft, remain the property of the original equipment manufacturer. The Air Force has limited ability to access these data, let alone take possession of these data and share them with third parties.

## *Upgrades, Modernization, and Service-Life Extension*

The average life cycle for an Air Force aircraft is more than 28 years.[4] Some Air Force systems (such as the C-130) live multiple lives—beginning first in service with the conventional Air Force and migrating eventually into the Air National Guard and also such commands as Air Force Special Operations Command (AFSOC), which heavily modify the base platform to tailor aircraft for their operational requirements.[5] For each of these evolutions in the life of a major weapon system such as an aircraft, the Air Force requires IP to upgrade, modernize, or extend the life of the system. The IP needed to complete these requirements often exceeds what exists within the OMIT or FFF data required and therefore requires the provision of an unlimited rights license that would allow the participation of third-party contractors in this activity. In the commercial aviation sector, there is a baseline of IP in the "flight worthiness data" required by FAA to be delivered and maintained for collective use.[6] There is also a custom and practice of using FFF data and a certain degree of modularity and interoperability with commercial systems to substitute components and modernize aircraft based on the FFF data provided as part of a commercial aircraft acquisition. However, current procurement law limits the Air Force's ability to use OEM IP (to the extent that it has possession of it) for these purposes, let alone share these data with third-party contractors, and there is no analogous reservoir of flight worthiness data the Air Force can use to support its aircraft either, especially after major modifications that might affect the flight profile of an aircraft.

## *Procurement of Additional Systems and Subsystems*

Over the life of an Air Force major weapon system, the service is likely to need additional systems or subsystems—whether to increase the size of its fleet, replace damaged or destroyed systems, or acquire capabilities that can be shared with additional parts of the service (such as AFSOC or the reserve components or other services). If the Air Force had sufficient IP about the system or subsystem, it could use competitive selection to choose a contractor from which to procure the additional systems or subsystems. Without adequate IP, the Air Force will be limited to procurement from the OEM. The Air Force may require additional IP than it typically receives

---

[4] Heritage Foundation, *An Assessment of U.S. Military Power: U.S. Air Force*, Washington, D.C.: Heritage Foundation, October 4, 2018.

[5] Camm et al., 2021.

[6] See Van Atta et al., 2017.

under a production contract, particularly for the integration of new subsystems into legacy aircraft.

### Replacement of Diminished Sources

Over the long life cycle of an aircraft, it has been the Air Force's experience that the designers and manufacturers of its systems and subsystems go through business cycles of their own. They may be acquired by a new parent company, sold in pieces to acquirers, or go out of business altogether. New manufacturing methods may also displace older ones necessary to maintain older systems; certain materials or parts may simply become unavailable from their original sources. Yet the Air Force must continue to operate and support its aircraft, notwithstanding these changes within its industrial base. This may require the Air Force to create new sources for parts or support, such as by reverse engineering those items or working with industry to create new sources of supply and service. Doing so requires the IP of the OEM— often in ways that far exceed the original licenses granted for that IP and the OEM's intent in sharing it.

## Two Related Sets of Problems: Legacy and New Systems

The Air Force faces two overlapping and related sets of challenges when developing strategies for IP for its major weapon systems. The first set of challenges arises during procurement, operation, and sustainment of legacy programs. The Air Force spends the majority of its O&S dollars on legacy systems, in absolute and percentage terms.[7] However, there may be less room to make new decisions with respect to IP on these programs, because the original EMD contracts for these legacy projects may have been awarded decades ago, reflecting the strategic, fiscal, and procurement policies of the time. If the Air Force wants to change the decisions it made about IP in the past, it typically faces a sole source provider that controls any additional IP that Air Force might want to access. The Air Force has limited leverage when negotiating with any sole source provider.

For new systems, a different set of challenges exists, based on the fact that design, prototyping, and EMD contracts have not yet been finalized. There are opportunities to realize great improvements, too, based on the bargaining power and leverage the Air Force has at the time of competitive award for an EMD contract—but the stakes are high for industry as well, based on the long revenue streams that lie ahead at the time of contract negotiation. If the Air Force wants to acquire more expansive IP rights so that it can create optionality in the future with respect to doing organic maintenance or using third-party contractors, OEMs will likely

---

[7] Michael Boito et al., *Expanding Operating and Support Cost Analysis for Major Programs During the DoD Acquisition Process: Legal Requirements, Current Practices, and Recommendations*, Santa Monica, Calif.: RAND Corporation, RR-2527-OSD, 2018.

attempt to raise their prices for IP to reflect the greater IP rights being sold and potential lost future revenue streams.

Box 2.1 summarizes illustrative opportunities to use IP in legacy and new systems.

**Box 2.1. Illustrative Opportunities for Use of Intellectual Property**

**Legacy Systems**
- Procurement of additional IP to facilitate organic support or depot-level maintenance
- Procurement of additional IP to facilitate contracting with third-party contractor for support or depot-level maintenance
- Procurement of additional IP to facilitate upgrades, modifications, and service-life extension for systems and subsystems
- Forensic auditing of legacy contracts to determine existing rights in IP
- Procurement of additional platforms

**New Systems**
- Choice of license for Air Force rights in technical data and rights in software
- Creation of contract deliverables to ensure delivery of IP
- Engagement in markings disputes to clarify Air Force IP rights during EMD contract performance
- Development of Air Force doctrine, operational concepts, training manuals, etc.
- Inclusion of new systems' intellectual data in Air Force business information systems (i.e., predictive analytics of maintenance and supply needs)

Ultimately, the Air Force leadership must prioritize which of these challenges it deems most important—and which to pursue—given finite resources in terms of money, acquisition personnel time, and the imperative to constructively work with OEMs and other industry partners to continue production and support for the service. Legacy systems may offer the greatest near-term potential for cost savings because of the money currently spent there; however, there may be the least contractual latitude with legacy systems. Conversely, new systems offer a greater opportunity for future cost savings per dollar of Air Force investment to get IP. But the share of future O&S costs will always be much smaller than that for legacy systems.

# 3. Effort Today to Yield Benefits Elsewhere in the Far Future

As noted in Chapter 2, the benefits to the Air Force of having technical data and data rights for a weapon system come mainly late in the life cycle of the system, when the Air Force operates and supports the weapon and potentially modifies it or seeks sources of parts or subsystems that are no longer available to support the system. If the Air Force waits until it needs technical data to acquire them, it finds itself facing a sole source with more leverage than the Air Force in any negotiation to acquire data or rights. But if the Air Force acts early enough to use competition to increase its own leverage—typically in the source selection for EMD—it must expend resources and effort decades before it will realize the benefits of this expenditure. When can the Air Force justify expenditures today that it will not benefit from for decades?

## Life-Cycle Planning with an Office of Management and Budget Discount Rate

This is a problem the Air Force faces whenever it conducts life-cycle planning. For example, the problem is essentially the same as that of spending resources on system design during EMD that will largely shape the military capabilities and O&S costs the Air Force realizes decades later. When facing such a problem, the Office of Management and Budget (OMB) offers a simple solution: use OMB's prescribed social discount rate to calculate the net present value (NPV) of future benefits and compare this NPV with the cost of the actions required early in a system's life cycle to realize the benefits.[1] If the NPV of the benefits exceeds that of the costs, make the investment. If not, use the scarce resources available in any budget year to do something else.

For example, suppose a new system takes ten years to design and build and then remains in the inventory for 30 years following initial operational capability. OMB's currently prescribed real long-term discount rate is 0.6 percent a year.[2] An annual flow of benefit of $1.00 over the system's 30 years of operations is worth $27.38 in real dollars today. That means that the Air Force should be willing to pay up to $27.38 in real dollars at the start of EMD for any feature of the system that will yield an annual flow of a dollar in benefits for the system.

---

[1] NPV is defined as the following, where A is the annual value of a flow, that flow will continue for T years, and the real discount rate is r: $A/(1+r) + A/(1+r)^2 + \ldots + A/(1+r)^T$.

[2] A "real" discount rate applies to flows of "real" dollars—current-year dollars that do not reflect future inflation. Office of Management and Budget, *Discount Rates for Cost-Effectiveness, Lease Purchase, and Related Analyses*, Circular No. A-94 Appendix C, Washington, D.C.: Office of the President, November 2017.

OMB's low discount rate means that the Air Force should be willing to pay a great deal for future benefits. For example, if effective access to technical data would allow the Air Force to reduce O&S costs by $10 million a year over the life of the weapon system in the example above, the Air Force should be willing to pay up to $257.9 million at the start of EMD to have access to those data.

## Discount Rate Suggested by Observed Air Force Behavior

In practice, observed Air Force decisions suggest that Air Force leaders are not willing to pay this much for benefits that accrue in the far future. No broad consensus exists on why this occurs. Possibilities include the following:

- The personnel making the decisions do not expect to be present when the benefits are realized. They cannot be held accountable for outcomes realized years after they have moved on. They prefer to commit the limited resources that the Air Force has during the budget year to efforts that yield benefits more rapidly.
- The budget available for investments in future capabilities and cost savings during the budget year is so small that personnel making these decisions believe they can spend all these limited funds during the budget year on investments likely to yield much greater rates of return than the OMB discount rate. Under these circumstances, they do not have a large enough budget to fund every investment that meets the OMB test. As a result, they focus investment on options that have higher rates of return. Doing this squeezes out the funds available to investments with lower rates of return. In effect, the decisionmakers apply a discount rate higher than the official OMB discount rate.
- The personnel making the decisions believe that some future benefits should have a higher priority than others when the Air Force compares alternative investments. Military capability gets the highest priority because, in a shooting war, that is what will most directly affect military success. O&S costs are less important because the Air Force can rely on Congress to appropriate the money needed for O&S, particularly in a shooting war.[3] So, to the extent that this argument explains behavior, these personnel might act as though they apply the OMB rate for military capabilities but apply a higher discount rate to O&S costs.

So, for example, suppose that arguments of this kind lead Air Force personnel to act as though they are applying a real annual discount rate of 10 percent. Under the first bullet, this might occur because the personnel believe they are less and less likely to be held accountable for the performance of a system, the further and further into the future that performance is realized and assessed. Under the second bullet, personnel could believe that they can invest every dollar

---

[3] For example, the Acting Chief Management Officer for DoD, Lisa Hershman, has strongly suggested that the warfighter is the most important customer of DoD and that it is better to support that customer with military "performance" than with "affordability." Quoted in Luis Martinez and Elizabeth McLaughlin, "How the Pentagon Has Saved $4.7 Billion in the Last Two Years," *ABC News*, January 26, 2019.

they have for investment during the budget year on projects that yield 10 percent or more. Under the third bullet, they might believe that improved military capability is several times more important than reduced O&S costs. Whatever the motivation, the NPV for an investment made in the first year of EMD drops from $27.38 to $9.43, or by 66 percent. These personnel are now willing to pay only about a third of what they were before to make investments that yield the same level of benefits in the far future.

This reduction in willingness to pay occurs for any type of investment that yields benefits in the far future. Early investment in technical data and data rights is only one example of such an investment. If senior Air Force leaders value benefits that occur in the far future in this light, they will tend to find other places to spend scarce investment dollars in the budget year, even if an investment in data and data rights could yield large benefits. To draw the interest of the leadership, the benefit must be large enough to overcome the fact that the Air Force must wait so long to get them.

## Whose Effective Discount Rate Is Higher—the Air Force or Its Contractors?

No empirical evidence is available to assess whether senior Air Force decisionmakers act as though a 10 percent discount rate is appropriate. But it is less important to know whether that rate is compatible with their behavior than to know whether Air Force leaders implicitly apply a discount rate higher than that which applies in the private sector.[4] OMB has determined that "a real discount rate of 7 percent . . . approximates the marginal pretax rate of return on an average investment in the private sector in recent years."[5] Suppose that an OEM would use a real discount rate close to this to value its technical data and rights to them. Suppose giving the Air Force effective access to its technical data cut an OEM's profits by a constant amount each year over the period that the Air Force operated and supported a weapon system. This might occur, for example, because giving the Air Force access to these data would allow the Air Force to turn to an alternative source for support if the OEM charges too much for that support. Given a 7-percent real discount rate, before giving the Air Force access to its technical data, the OEM would require a payment of at least $6.31 at the beginning of EMD for every loss of an annual flow of $1.00 in profit over the operational life of the system.

Suppose that when the OEM grants the Air Force access to its technical data and right to use them, the Air Force benefits a dollar's worth for every $1.00 of profit that the OEM loses.

---

[4] In its Tension Point Paper 1, "Different Business Models in Government and Industry Result in Different Objectives," the Section 813 Panel Report frames this comparison in terms of basic differences in business model between the government and a contractor. We focus on the relative cost of capital to highlight a fundamental difference between the government and any contractor when they make decisions with consequences many years in the future.

[5] Office of Management and Budget, *Guidelines and Discount Rates for Benefit-Cost Analysis of Federal Programs*, Circular A-94, Transmittal Memo No. 64, Washington, D.C.: Office of the President, undated.

In effect, the access to data transfers profits from the OEM into Air Force hands by reducing the amount that the Air Force will have to pay for O&S in the future. If senior Air Force decisionmakers act as though their real discount rate is lower than 7 percent a year, the Air Force can benefit more from getting access to technical data than the OEM gives up when it provides such access. If Air Force decisionmakers used the official OMB discount rate of 0.6 percent a year, they would be willing to pay the OEM $27.38 for every $1.00-per-year flow of benefit that access to technical data would give the Air Force. OEM officials would be willing to provide access as long as the Air Force paid them more than $12.41 for such a flow. The difference between $12.41 and $27.38 potentially sets the stage for substantial mutual gains from trade.

Table 3.1 illustrates the magnitude of potential mutual gains available from moving access to technical data when the Air Force and OEM have differing real discount rates. The table lists real discount rates for the OEM across the top of the table and those for the Air Force down the left-hand side. For each real discount rate, the table displays the NPV of an annual gain of $1.00 over the 30-year life of a new weapon system—for the OEM across the top and for the Air Force along the left-hand side. So, for example, if the OEM had a real discount rate of 1 percent, gaining $1.00 a year for 30 years would pay the OEM an NPV of $25.81. If the government had the OMB-prescribed real discount rate of 0.6 percent, gaining $1.00 a year for 30 years would pay the Air Force an NPV of $27.38. For any pair of discount rates, the body of the table shows the potential gain that the OEM and the Air Force could share if that pair of discounts applied and the Air Force gained $1.00 for every $1.00 the OEM lost. If the OEM rate were 1 percent and the Air Force rate were 0.6 percent, $1.57 would be available for the Air Force and the OEM to split through negotiation. The higher the OEM rate relative to the Air Force rate, the more is available to split. In our example above, we assumed rates of 7 percent and 0.6 percent, yielding a net potential gain of $14.97 for every $1.00 of annual gain to the Air Force and loss to the OEM.

**Table 3.1. Potential Mutual Gain from Moving Access to Technical Data from the Original Equipment Manufacturer to the Air Force**

| | | | Annual Real OEM Discount Rate | | | | | |
|---|---|---|---|---|---|---|---|---|
| | | | 0.006 | 0.01 | 0.05 | 0.07 | 0.1 | 0.15 |
| | | Total OEM gain | 27.38 | 25.81 | 15.37 | 12.41 | 9.43 | 6.57 |
| | | Total AF gain | | | | | | |
| Annual Real Air Force Discount Rate | 0.006 | 27.38 | 0.00 | 1.57 | 12.01 | 14.97 | 17.95 | 20.81 |
| | 0.01 | 25.81 | −1.57 | 0.00 | 10.44 | 13.40 | 16.38 | 19.24 |
| | 0.05 | 15.37 | −12.01 | −10.44 | 0.00 | 2.96 | 5.95 | 8.81 |
| | 0.07 | 12.41 | −14.97 | −13.40 | −2.96 | 0.00 | 2.98 | 5.84 |
| | 0.1 | 9.43 | −17.95 | −16.38 | −5.95 | −2.98 | 0.00 | 2.86 |
| | 0.15 | 6.57 | −20.81 | −19.24 | −8.81 | −5.84 | −2.86 | 0.00 |

12

The price of the transfer of data will depend on the relative leverage of the parties. The more leverage the Air Force has, the closer the price will be to $6.31. The Air Force has maximum leverage when offerors are competing with one another in the source selection for EMD. The more leverage the OEM has, the closer the price will be to $25.79. For example, if the Air Force waits until it needs access to technical data, the sole-source OEM will have close to maximum leverage and will be able to extract most of the gains from trade from the Air Force.

This is where a fundamental difference in financing between the OEM and the Air Force becomes important. The OEM has direct access to a capital market. The real discount rate of 7 percent reflects what money costs the OEM in the commercial financial market. Whatever money the OEM can extract from the Air Force when it transfers data and data rights to the Air Force, it can apply immediately to the OEM's cash needs, reducing the OEM's need to borrow or apply internal equity funds by a dollar for every dollar it receives from the Air Force. So whenever the OEM makes decisions to move money between this budget year and future years, it has immediate access to a financial mechanism it can use to do that. The Air Force has no comparable mechanism. Even if OMB states that the real discount rate for investment funds is 0.6 percent a year, no one in the Air Force can access a capital market that would provide funds at that rate. The Air Force can only commit during any budget year the funds that Congress grants to the Air Force in that year.

This will become important below, because almost everything the Air Force must do to get better access to technical data and data rights in the future requires an investment of effort and monetary resources today. Simply to frame the importance of getting access to data and data rights, the Air Force must commit effort during the definition of requirements and the acquisition strategy for a new system. To reflect data and data rights in a source selection, the Air Force must commit resources to negotiation and evaluation. To verify that the Air Force gets the data and data rights it has contracted for, a program office must commit the scarce time and effort of the program manager and program staff. In each situation, the Air Force is committing effort and resources today to get benefits in the far future.

In each situation, the OEM is doing the same thing. It commits resources and effort to industry days and negotiation over the terms of the request for proposal. It frames a proposal on data and data rights and negotiates that through source selection. It then creates data and negotiates the effective rights attached to these data through the course of EMD and beyond. At each step in this process, the OEM and government are applying effective discount rates to decide how much effort is worthwhile to protect or secure benefits for themselves in the far future. We can presume that the OEM's direct access to a capital market leads it to use a real discount rate of about 7 percent a year to make these judgments. If the Air Force is applying a rate near the OMB rate, it is willing to apply a great deal more effort than the OEM—up to about four times ($25.79 versus $6.31) for each flow of $1.00 per year over the system's operational life that is in play. If, on the other hand, the Air Force acts as though the annual real opportunity cost of its efforts and resources is closer to 10 percent, it will be willing to apply less effort and resources than the OEM for equivalent future flows of benefits—just 58 percent as much ($3.63/$6.31) in our example.

## Summary

In the realm of technical data and data rights, both the Air Force and the OEM must invest now to achieve benefits in the far future. How much they are each willing to invest depends on the effective real discount rates they apply to decide how to allocate resources over time. If the Air Force applies a lower rate than the OEM, room for potential mutual gains from trade exists for the technical data and data rates. And, at least as important, the Air Force is willing to apply more effort to getting what it wants in the future than the OEM is. If, on the other hand, the Air Force applies a higher rate than the OEM, room for potential mutual gains is not likely to exist. And the OEM is willing to apply more effort than the Air Force to narrow any data and data rights that may be due to the Air Force. If the senior Air Force leadership concludes that the real discount rate relevant to its decisions about policy on technical data and data rights is lower than the OEM's, the chapters that follow offer advice that the leaders can act on. If not, it is unlikely that these leaders will be able to justify the broad program of effort the Air Force will need to pursue to improve its access to technical data and data rights.

# 4. What the Air Force Must Do to Acquire the Technical Data and Data Rights It Needs

During the Cold War, the Air Force relied heavily on in-house support of its weapon systems after enough experience had accumulated during an interim contract support period to write technical orders that government personnel could use to maintain systems. The Air Force acquired extensive technical data and data rights to support this approach to system sustainment. It maintained technical expertise in-house to support this approach. Beginning in the 1990s, the Air Force shifted steadily toward more and more contractor support.[1] The Air Force steadily built this shift into new systems by planning from the beginning to rely on contract support. If support would remain in the hands of the OEM, the Air Force's need for technical data and data rights dropped considerably. As Air Force emphasis on getting technical data and data rights fell, Air Force personnel lost their understanding of the importance of those elements. Emphasis shifted to other issues.

The Air Force has decided that it can no longer rely so heavily on OEM sustainment over the lives of its systems, both because of the costs for such an approach and because of the situations where the Air Force wants to use its own personnel or a third-party contractor to maintain, modify, upgrade, or operate systems in ways that require the OEM's IP.[2] It must now rebuild its ability to acquire technical data and data rights in an environment where that standard approach has been to leave these data and data rights with the OEM. This chapter reviews a set of capabilities that would enhance the Air Force's ability to get the data and data rights it wants. The discussion traces a system from the development of its requirements through source selection to EMD, production, and fielding. The Air Force needs new capabilities at multiple stages to any new system's life cycle.[3]

---

[1] For details on this trend, see Thomas Light et al., *Understanding Changes in U.S. Air Force Aircraft Depot-Level Reparable Costs over Time*, Santa Monica, Calif.: RAND Corporation, RR-2518-AF, 2018.

[2] These issues have been discussed in many recent studies of the sustainment of DoD systems. See Michael Kennedy et al., *USAF Aircraft Operating and Support Cost Growth*, Santa Monica, Calif.: RAND Corporation, RR-813-AF, 2014; Mark A. Lorell, Robert S. Leonard, and Abby Doll, *Extreme Cost Growth: Themes from Six U.S. Air Force Major Defense Acquisition Programs*, Santa Monica, Calif.: RAND Corporation, RR-630-AF, 2015; Michael Boito, Thomas Light, Patrick Mills, and Laura H. Baldwin, *Managing U.S. Air Force Aircraft Operating and Support Costs: Insights from Recent RAND Analysis and Opportunities for the Future,* Santa Monica, Calif.: RAND Corporation, RR-1077-AF, 2016; Boito, Conley, et al., 2018; Light et al., 2018.

[3] The material reported here draws on official documents that describe appropriate management of technical data in a government setting, interviews with personnel in a variety of Air Force and joint program offices, and secondary literature on activity in these offices.

The discussion below assumes that the Air Force has taken measures to ensure that the personnel responsible for making decisions related to technical data and data rights understand the importance of data and rights and have access to expertise on these topics that they can use to support their decisionmaking. Based on interviews with Air Force personnel and research on IP issues affecting current programs, such understanding and access to expertise are quite limited in the Air Force today.

## Requirements

The Joint Capabilities Integration and Development System (JCIDS) process transforms warfighter priorities into capability gaps, some of which can be filled with new Air Force material systems.[4] JCIDS prescribes a set of capability attributes that this process must address. Among those that JCIDS considers "critical or essential to the development of an effective military capability" is sustainment. JCIDS calls for the definition of requirements for sustainment in terms of a key performance parameter (KPP) and affiliated logistics factors that explain how reliable, available, maintainable, and affordable a new system should be to meet its capability requirements.[5] In particular, the KPP defines a high-level trade space bounded by threshold and feasibility levels of material and operational availability.[6] The JCIDS system associates a set of indentured measures of reliability, maintainability, and O&S cost with these threshold and feasibility levels.

These statements of requirements are relevant in our setting in two ways. First, they set the stage for defining the performance factors that the source selection for a material system will use to judge offers that address the capability gap that JCIDS has identified. The acquisition community must link the factors it includes in any source selection to performance parameters documented in JCIDS capability requirements documents. Second, they potentially set the stage for identifying what technical data and data rights the Air Force will need to successfully implement any sustainment strategy it might plan for a new system once it is fielded.[7]

---

[4] Office of the Joint Staff, J-8, *Manual for the Operation of the Joint Capabilities Integration and Development System*, Washington, D.C.: Department of Defense, August 31, 2018.

[5] Office of the Deputy Assistant Secretary of Defense for Systems Engineering, *Reliability, Availability, Maintainability, and Cost (RAM-C) Rationale Report Outline Guidance*, Version 1.0, Washington, D.C.: Office the Secretary of Defense, February 28, 2017.

[6] "Materiel" availability is the proportion of a fleet that is available. "Operational" availability is the proportion of times that a set of systems are available to execute a particular mission.

[7] The Section 813 Panel Report discusses these points in more detail in its Tension Point Paper 12, "Are Existing Rights Sufficient for Maintenance and Sustainment?"

In principle, acquisition logisticians should be present in the JCIDS process to validate the availability of KPP for sustainment and its associated reliability, maintainability, and O&S cost factors. IP specialists should be present as well to begin planning the IP strategy for the new system. Presumably, these logisticians and IP specialists collaborate to identify the data and data rights relevant to any logistics plan.

This does not occur when the sustainment plan calls for the OEM to provide life-cycle sustainment. Once that decision was made beginning in the 1990s, the government had only limited needs for data. The Defense Federal Acquisition Regulation Supplement (DFARS) is clear that the Air Force should acquire no more data and data rights than it needs.[8] In recent years, that has led Air Force and OEM personnel to de-emphasize the relevance of technical data and data rights issues during source selection and contract formation.

Suppose that the JCIDS process included a requirement that the Air Force maintain flexibility about the sustainment approach it would use over a weapon system's life cycle. That would suggest that the sustainment approach adopted initially need not be expected to remain in place through the life cycle. A requirement for such flexibility could provide a justification for the Air Force to acquire more technical data and data rights than dictated by the sustainment approach initially planned during EMD and initial fielding. Such a requirement would have dictated a sustainment KPP and associated logistics factors that were robust in the face of alternative potential sustainment plans. But a variety of sustainment choices would be available only if IP specialists anticipated the desirability of this variety and made it possible with an IP strategy that acquired a more robust set of technical data and data rights during requirements determination.

Note that bringing more complete consideration of technical data and data rights forward into the requirements determination process increases the length of time between (1) the investment of resources and leadership effort in data and data rights and (2) the realization of benefits for the Air Force from these data and data rights. That is worthwhile if the effective discount rate the Air Force leadership uses to allocate resources across time is low enough and, in particular, lower than the discount rate that the OEM applies. If the Air Force discount rate is low enough, mutual opportunities for gain will exist for the OEM to transfer technical data and data rights to the Air Force on reasonable terms. And the Air Force will have the effective will to commit resources early to counter OEM pushback as the requirements development process refines requirements for technical data and data rights. If the Air Force discount rate is not low enough, mutual opportunities for gains will not exist. And the OEM will be willing to commit more resources than the Air Force to shaping technical data and data rights during the requirements development phase.

---

[8] DFARS 227.7103-1 and 227.7203-1.

## Acquisition Strategy, Market Research, Source Selection, and Contract Formation

These phases of a weapon system's life cycle progressively refine the Air Force's capability requirements into detailed and precise language that the Air Force can use to induce an OEM to meet those capability requirements. The refinement of requirements for specific technical data and data rights is one element of a much broader effort that should not be separated from that broader effort. This is where opportunities exist for structured interaction between the Air Force and potential offerors to clarify the Air Force's life-cycle sustainment plan and the technical data and data rights it believes it will need to realize that plan. That interaction can potentially draw on contractor expertise to improve that sustainment plan in ways that help the Air Force learn how to take full advantage of the feasible options available for sustainment. It can also clarify where differences exist in the Air Force's and contractors' understanding of what fundamental concepts like OMIT cover, what FFF data the Air Force expects and at what level of indenture down into the supply chain, how government purpose rights will be defined in practice, what markings will be acceptable on drawings, and so on.[9]

In principle, the government and industry can cooperate in broader efforts to build a common understanding of what these central concepts mean. As a practical matter, a successful source selection depends on achieving a common understanding of these things for each program that the Air Force pursues. Formation of an effective contract—a contract that facilitates the successful execution of EMD—similarly depends on achieving a common understanding of these things for each program that the Air Force pursues. Perhaps the interaction that occurs in these phases for each system acquisition can help fuel broader efforts at common understanding, but each program acquisition will have to proceed with whatever level of understanding it can achieve, even as efforts at broader understanding proceed in the background.

Once again, it is worth placing these activities in temporal context. They occur a bit closer to the time when well-defined technical data and data rights can yield benefits for the Air Force than the requirements development process is. So it is potentially easier for the Air Force to justify investing effort in technical data and data rights here than earlier. But well-defined requirements open the door for effective consideration of data and data rights during these phases. Without more complete preparation during requirements development than we have now, the Air Force will have difficulty justifying the data and data rights it wants to highlight here. In the same way, more complete preparation here clears the way for more successful EMD activities.

---

[9] The Section 813 Panel Report discusses these points in more detail in its Tension Point Papers 15, "Operation, Maintenance, Installation, and Training (OMIT) Data Versus Detailed Manufacturing or Process Data (DMPD)," and 17, "Poor Data Item Description (DID) Alignment with Statutory/Regulatory Categories Form, Fit and Function (FFF) and Operation, Maintenance, Installation or Training (OMIT)."

But all these activities remain decades away from the operational phase when the Air Force can benefit most from improved technical data and data rights. The Air Force's effective discount rate is almost as important here as it is during requirements development. A low discount rate opens the way for mutual gains with the OEM and can motivate the Air Force to commit the resources it will need to address technical data and data rights effectively during these phases. Effective interaction with potential offerors during these phases depends on the Air Force showing the will to commit at least the same effort that the offerors will. The lower the Air Force's effective discount rate, the more productive these phases will be to the Air Force's effort to prepare for the future. If the Air Force's effective discount rate exceeds that of the offerors, potential gains from trade may not exist. And the level of commitment that offerors bring to interactions with the Air Force during these phases could overwhelm the Air Force.

This reasoning raises special concerns about efforts to bring specially negotiated license rights (SNLRs) into the IP strategy for technical data. In principle, such rights offer the opportunity to tailor the terms of any acquisition in pursuit of mutual gains. More generic terms are likely to protect Air Force interests if the Air Force is not willing to bring as much effort and skill to negotiations as the offerors do. SNLRs are a classic example of how to invest up front to achieve benefits in the future. The Air Force should be cautious about pursuing SNLRs unless the effective discount rate of senior Air Force leaders is low enough, relative to that of their contractor counterparts, for the Air Force to invest the effort required to protect and project Air Force interests as it negotiates tailored language with its contractor counterparts.[10]

## Engineering and Manufacturing Development Contract Execution

EMD raises two issues relevant to technical data and data rights in the Air Force. First, the Air Force must assure that contractors deliver the technical data called for in contract data requirements lists (CDRLs) in a form compatible with agreements about data rights stated in the contract. The Air Force must take delivery in the form agreed on. And the Air Force must ensure that the data delivered are properly marked to reflect the Air Force's rights. Recent experience has shown that Air Force program offices are often not well informed about their responsibilities and their rights. And these responsibilities account for only a portion of what they must achieve with limited resources. Contractors can take advantage of this lack of knowledge and resources to deliver less than what was agreed on earlier.[11]

---

[10] Multiple Tension Point Papers in the Section 813 Panel Report strongly endorse use of SNLRs to tailor agreements on data and data rights. They do not raise the point discussed here. See, for example, Tension Point Papers 5, "Intellectual Property (IP) Valuation"; 8, "Is Source of Funding the Best Way to Determine Rights to Technical Data?"; 16, "Rigid Intellectual Property (IP) Requirements versus Flexible Arrangements."

[11] The Section 813 Panel Report discusses this issue in Tension Point Paper 22, "Lack of Trained Personnel."

Second, EMD is an inherently uncertain activity in which the Air Force and its contractors transform broad ideas about potential new capabilities into a specific design that can be manufactured, operated, and supported in a predictable manner. Surprises must be expected. When they occur, they typically force the Air Force and its contractors to revisit the agreements about performance, cost, and schedule that they made and documented in the initial EMD contract and EMD baseline. Surprises that force developers to increase the cost associated with completion, if serious enough, draw the attention of Congress in the form of Nunn-McCurdy breaches.[12] The potential for such breaches and similar shortfalls of other kinds gives contractors an opportunity to revisit earlier agreements about technical data and data rights. Contractors can offer to absorb cost increases or pay to keep development on schedule in return for concessions on technical data and data rights. Contractors can deliver fewer technical data or deliver them with more restricted government rights.

Both issues involve circumstances in which Air Force and contractor personnel make decisions about how many resources and how much effort to continue to commit to the acquisition of technical data and data rights documented in the initial contract. The Air Force faces these circumstances with limited resources and no direct access to a capital market. In these circumstances, it must ask again whether resources are worth committing today in exchange for benefits far in the future. If the effective Air Force discount rate is lower than that of a contractor, it is easier for a program office to commit resources today than it is for a contractor. If not, the contractor has the advantage.

With a low discount rate, relative to that of its contractors, a program office can explain to senior Air Force leaders and Congress why it invested effort today and even potentially accepted a breach in cost or schedule to protect the benefits that technical data and data rights can yield in the far future. Air Force lawyers can explain why they held their ground and forced a dispute into court to get a court judgment that the Air Force can use to support its claims to technical data and data rights in the future and why that was a good thing for the Air Force, even though going to court imposed costs and induced delays in a particular program. With a high discount rate, relative to that of its contractors, a program office can face a contractor prepared to commit more resources and more highly skilled personnel to a dispute than the Air Force is and to wait the Air Force out. This situation would be compatible with greater use of lawyers and IP professionals in day-to-day EMD activities by contractors than by the Air Force. A relatively low discount rate allows the contractor to be more patient than the Air Force as disputes go forward and development slows.

---

[12] The Nunn-McCurdy Act of 1983 (10 U.S.C. 2433) requires that DoD notify Congress when cost rises significantly above the planned baseline cost in a major defense acquisition program. For details and a useful discussion, see Moshe Schwartz and Charles V. O'Connor, *The Nunn-McCurdy Act: Background, Analysis, and Issues for Congress*, Washington, D.C.: Congressional Research Service, R41293, May 12, 2016.

## Operations and Support Contract Execution

The Air Force comes into the O&S phase with the technical data and data rights that have been specified in the initial contract and protected and delivered during EMD. The Air Force presumably has basic technical data and data rights relevant to the baseline weapon system in hand by the time O&S starts.

But as a weapon system ages, upgrades, life extension, and modernization almost always occur. The development elements of these activities raise all the same issues discussed above regarding EMD. That is, even if basic technical data and data rights are locked in for the basic weapon system for the duration of its life, any changes open opportunities to acquire more complete sets of data and rights for the elements added. In these situations, the difference in discount rate between the Air Force and its contractors becomes less important, because a much smaller delay exists between the investment in a change—and the data and rights associated with it—and the realization of improved O&S services associated with the change and the data and rights that accompany it. That may suggest that the Air Force leadership will be more willing to commit the resources and effort required to get selected data and data rights once operations have begun. Part of the effort, of course, includes ensuring that the program offices responsible for upgrades, life extension, and modernization understand the value of data and data rights.

Upgrades, life extension, and modernization will be easier if the program office has enough data and data rights on the baseline weapon system to facilitate integration engineering and testing.[13] To ensure that such data and rights will be available when the time comes, the initial contract and EMD effort that produced the baseline weapon systems must address these data and rights in CDRLs, protect them through EMD, and ensure appropriate delivery. Data relevant to integration and testing are likely to extend beyond the scope of data normally associated with OMIT and may require FFF data at a relatively deep level of indenture to diagnose problems that arise during integration and testing. The Air Force has the best leverage to act to ensure access to such data and rights in the definition of CDRLs in the EMD contract, decades before a program office will need them during the O&S phase.

As a system continues to age, it increasingly encounters circumstances in which the sources that originally provided subsystems and parts are no longer available to replace these subsystems and parts when they reach the ends of their useful lives. The Air Force can rely on its OEM to maintain access to such sources in its supply chain. If it wants the option of relying on an organic organization or on a contractor organization other than the OEM, it will need access to and rights to use FFF data defined in enough detail to allow the identification and validation of new sources. Again, the best place for the Air Force to ensure access to such data and rights is during the definition of CDRLs in the EMD contract.

---

[13] The Section 813 Panel Report discusses this issue in Tension Point Paper 11, "Authorized Release and Use of Limited Rights Technical Data."

## Summary

In sum, technical data raise issues for a system program office throughout the life cycle of any system. Table 4.1 summarizes the tasks discussed above that a system program office should execute routinely and repeatedly to ensure that the Air Force acquires and maintains appropriate access to technical data and data rights.[14] The first column identifies selected broad tasks that a system program office conducts over the course of a system program life cycle. The second column lists the tasks associated with government access to technical data that a program office should carry out in each of these phases.

**Table 4.1. System Program Office Tasks Associated with Access to Technical Data**

| Broad Program Tasks | Tasks Associated with Access to Technical Data |
|---|---|
| Acquisition strategy | • Assess data the Air Force needs |
| Solicitation/request for proposal (RFP) development | • Clarify data needs in specific CDRLs<br>• Use DFARS clauses to preserve flexibility in contract to revisit data issues |
| RFP response evaluation and source selection | • Review and challenge contractor assertions about rights |
| Post contract award and early contract | • Ensure timely delivery of data in CDRLs<br>• Review, challenge markings<br>• Document basis for assertions, markings, and funding split that Air Force accepts<br>• Preserve data on assertions, markings for future use |
| Operations and sustainment | • Preserve skills to discipline contractor over the life of weapon system<br>• Use leverage creatively to motivate contractor |

From the perspective of the formal requirements and acquisition processes in DoD, these tasks effectively begin in the technology maturation and risk reduction phase with requirements for government access to technical data defined in the capability development document.[15] The EMD source selection translates these requirements into acquisition requirements that become the basis for deliverables in the EMD contract. The government executes this contract during the EMD phase, which sets the stage for delivery of all technical data identified in the EMD contract and government review of all data delivered. Data delivered and the rights attached to them then frame the government's options during the production and deployment phase and O&S phase of the formal acquisition process.

---

[14] For more detail on these tasks, see Camm et al., 2021.

[15] For a simple overview of these DoD processes, see Figure 1, "Illustration of the Interaction Between the Capability Requirements Process and the Acquisition Process," in Department of Defense Instruction 5000.02, *Operation of the Defense Acquisition System*, Washington, D.C.: U.S. Department of Defense, Office of the Under Secretary for Acquisition, Technology, and Logistics, January 7, 2015, p. 6.

# 5. What Can Be Done to Define and Then Sustain a Longer-Term Perspective?

Chapter 4 points to the potential for seeking greater access to technical data and data rights on many fronts. To be successful, Air Force efforts must be coordinated. For example, unless efforts are taken early in a program's life cycle, the Air Force's ability to act later can be quite limited. But taking action early is not enough. Benefits will not accrue to that early action unless the Air Force follows through later in the life of a program. The breadth of this challenge suggests that the senior leadership of the Air Force should use formal change management techniques to ensure that actions throughout the Air Force are aligned to one another, across functions and across time.

This chapter explains what formal change management entails. It offers specific examples of ways the Air Force can enhance its access to technical data and data rights through formal change management. And it identifies which parts of the Air Force should have responsibility for specific parts of an Air Force–wide formal change management effort.

## Using Formal Change Management to Implement New Arrangements

The quality management movement that came to the United States from Japan in the 1980s and has grown and prospered ever since brought with it an approach to formal change management that large, complex public- and private-sector organizations have used successfully to implement a wide range of changes.[1] The approach has been studied and documented in publications aimed at academic, technical business, and broader popular audiences.[2] John Kotter became especially well known in the 1990s for the conclusions he drew from an extensive practice helping large

---

[1] For more on the link between the quality management movement and current U.S. thinking about formal change management, see Frank Camm, "Adapting Best Commercial Practices to Defense," in Stuart E. Johnson, Martin C. Libicki, and Gregory F. Treverton, eds., *New Challenges, New Tools for Defense Decisionmaking*, Santa Monica, Calif.: RAND Corporation, MR-1576-RC, 2003, pp. 211–246.

[2] For literature reviews of academic and technical business publications likely to be useful in an Air Force setting, see Gail L. Zellman et al., "Implementing Policy Change in Large Organizations," in Bernard D. Rostker et al., *Sexual Orientation and U.S. Military Personnel Policy: Options and Assessment*, Santa Monica, Calif.: RAND Corporation, MR-323-OSD, 1993, pp. 368–394; and Cynthia R. Cook et al., "Implementation," in Bernard D. Rostker et al., *Sexual Orientation and U.S. Military Personnel Policy: An Update of RAND's 1993 Study*, Santa Monica, Calif.: RAND Corporation, MG-1056, 2010, pp. 371–388.

corporations implement change.[3] This section draws on this literature to broadly summarize the elements of formal change management that they have emphasized.[4] It then suggests specific initiatives the Air Force might consider in applying these elements to policy efforts on technical data and data rights.

### *Recent Formal Change Management in the United States*

Formal change management can be described in many ways. For simplicity, we break the change process into three elements—preparation, execution, and sustainment. Preparation ensures that everything needed to execute a discrete, well-defined change is in place before it begins. Execution ensures that the change takes place as planned and is adjusted as needed in response to information gathered during the change. Sustainment ensures that the positive effects of the change persist after senior focus on the change moves on to other high-priority activities. It ensures that the change is institutionalized in a new set of standard procedures.

Some authors refer to these as steps or phases. But it is important to remember that in the presence of continuous improvement, a notion that continues to dominate current thinking, it is best to think of change as a series of increments. Each builds on the last by monitoring how well the previous increment worked and adjusting the next to benefit from what can be learned from the empirical performance of the last. In such a setting, at any point in time, a large organization will be planning some incremental changes, executing others, and sustaining or institutionalizing others. And information will be flowing among all these efforts all the time. So even though we will describe these pieces as though they occur sequentially, in fact, the pieces often have a great deal of overlap, with feedback moving back and forth across them all the time.

This dynamic view of change seems well suited to the Air Force context. The senior leadership at the service, major command, and wing level turns over on a regular schedule. Yet the Air Force can look out decades in the future and introduce new programs and weapon systems that no one leadership team could complete on its own watch. The Air Force is used to structuring large changes so that leaders fall in on a change effort in progress and hand it off to the next team at the end of their tour.

The management of an aircraft program offers a simple way to see this. At any point in time, a complex aircraft program has multiple blocks at various points in their life cycles. Future blocks are still assembling technologies that can make new capabilities feasible. Other blocks not so far in the future are in development and testing. Still others approaching operational release

---

[3] John P. Kotter, *Leading Change*, Boston: Harvard Business School Press, 1996.

[4] For more detail on these elements, see "Implementation," Chapter 7, in Frank Camm et al., *Charting the Course for a New Air Force Inspection System*, Santa Monica, Calif.: RAND Corporation, TR-1291-AF, 2013. The discussion here draws heavily on this source.

are in production. And still others are in the fleet, where the Air Force monitors their performance and adjusts support plans as they age. That is how the formal change management approach that we describe here works as well. In fact, the advocates of that approach could have learned a great deal from how the Air Force (and the rest of DoD) has managed weapon system programs for decades.[5]

Within each element, a common set of activities occurs. The exact nature of these activities changes as a change process moves from preparation to execution to sustainment. Key activities include the following:

- *Alignment.* All work must be aligned to the high-level organization goals of the change effort. This is important to us because, to give technical data and data rights more emphasis in the Air Force, players in many different parts of the Air Force must coordinate their actions over long periods.
- *Resourcing.* Resourcing takes two forms. One is the provision of institutional and financial support to pay for each of the activities associated with change. The second is an effort to relieve personnel involved in the change of other duties so that they can contribute effectively.
- *Training.* Training takes two forms. The first is training that ensures that relevant personnel know how to execute their responsibilities once the new emphasis on technical data and data rights is in place. The second is training in formal change management itself and its coordination with the day-to-day activities of an organization that must continue to be performed as change proceeds.
- *Monitoring and measurement.* Monitoring typically requires a new governance structure that ensures that information relevant to the progress of ongoing change is collected and reported to those who need it to manage ongoing change. Measurement requires the identification of appropriate quantitative and qualitative metrics that monitors can use to report information on the progress of change. Change management typically requires more information than day-to-day management does to ensure that the organization can learn from the experiments implicit in its incremental efforts.
- *Motivation.* Motivation convinces every person who must change her or his behavior for change to succeed that she or he will be better off if change succeeds than if it does not.
- *Communication.* Communication moves information in all directions throughout the organization undergoing change. It moves information on what the senior leadership wants, on how incremental change efforts are proceeding, on challenges encountered and methods for addressing them as they are devised, on the consequences of ongoing change for all involved, and on how the operation of the organization is adjusting in a way that ensures that coordination across the organization continues even as standard procedures change.

---

[5] The Air Force Planning, Programming, Budgeting, and Execution System has a similar and completely analogous structure. The budget generated by each cycle is a discrete, incremental product of the longer-term process. At any one point in time, elements of planning, programming, budgeting, and execution activities in different cycles proceed simultaneously, constantly informing one another.

- *Building on incremental successes.* Incremental efforts test new ideas and adjust them to ensure that they yield the results desired. They allow local failures that generate useful information for broader change efforts without threatening the performance of the organization as a whole. Incremental success gives the senior leadership confidence that change is worth the effort in terms of resources and general operational disruption.
- *Consolidation of successes.* To be successful, change typically requires the focused attention of the leadership. As success occurs and spreads until it touches all parts of the organization, the leadership can turn its focus elsewhere and rely on new procedures to ensure that the benefits of change are institutionalized.

Table 5.1 summarizes these activities and describes briefly how they evolve as a formal change effort moves from preparation to execution to sustainment. Experience has taught us that these activities are easy to describe and hard to conduct successfully, particularly over extended periods as personnel involved in the change effort turn over.

**Table 5.1. Elements of Formal Change Management Emphasized in Recent Publications**

| Activity | Prepare | Execute | Sustain |
|---|---|---|---|
| Align | Create a coalition of all who must change | Link efforts to high-level goals; stay on target | Align new activities to standard procedures |
| Resource | Provide staff time, resources | Provide staff time, resources | Provide staff time, resources |
| Train | Train on specific change Train on change management | Adjust based on experience to date | Adjust based on experience to date |
| Measure | Devise monitoring regime, metrics linked to goals | Monitor progress, adjust plan | Monitor progress, report success |
| Motivate | State a clear, compelling need to change | Reward success | Reward success |
| Build on incremental successes | Devise an executable plan | Build on success | Expand to completion |
| Communicate | State the need for change and the plan to achieve it | Share data on progress, success, challenges | Share data on evidence of success and its consequences |
| Consolidate successes | | Consolidate successes as they accrue | Anchor successes in standard procedures |

## Specific Activities Relevant to Technical Data and Data Rights

Broadly speaking, if the senior leadership wants to improve Air Force access to technical data and data rights, it needs to mobilize and sustain a virtuous cycle that builds on demonstrated success to encourage continuing commitment to further change, as exemplified in the following:[6]

- Explain why technical data and data rights are important to the Air Force. For example, they can enhance long-term O&S performance and cost; simplify upgrading, modernization, and life extension for aging systems; or simplify Air Force access to sources of parts and subsystems as the industrial supply infrastructure ages.
- Set a short list of, say, four to six specific goals the leadership wants to achieve over a stated period—perhaps one to two years. State clearly how these help the Air Force improve its access to more technical data and data rights. State clearly who is responsible for the changes and what the consequences of success and failure are for the Air Force and for those held responsible. Clarify their responsibility in terms that are compatible with the professional standards of their function communities.
- Identify clear metrics—quantitative or qualitative—to track performance against these goals. Report these metrics regularly throughout the Air Force to maintain awareness and commitment.[7]
- Use all mechanisms of motivation available to the Air Force—personal and organizational recognition, promotion, expanded responsibilities, access to desirable assignments and training, and even money if possible—to reward individuals and organizations that succeed in achieving change at the expense of those that do not.[8]
- Use cumulative evidence of success to maintain senior leadership support as the senior leadership turns over and to promote the legitimacy of change to all those being asked to change their behavior.

These are challenging activities. Often, they fail in large, complex organizations in the commercial sector. That is true mainly because they involve many moving parts that must all work together to achieve success. To succeed, the senior Air Force leadership must invest today in multiple communities to achieve Air Force–wide improvements in performance and cost in the far future. These include the acquisition, legal, personnel, and technical communities of the Air Force responsible for requirements development; acquisition strategy, market research, source selection, and contract formation; EMD; and O&S. In each case, one community is

---

[6] For more detail in different defense settings, see Frank Camm et al., *Implementing Proactive Environmental Management: Lessons Learned from Best Commercial Practice*, Santa Monica, Calif.: RAND Corporation, MR-1371-OSD, 2001; Nancy Young Moore et al., *Implementing Best Purchasing and Supply Management Practices: Lessons from Innovative Commercial Firms*, Santa Monica, Calif.: RAND Corporation, DB-334-AF, 2002; Camm, 2003.

A "virtuous cycle" is a series of activities that together create a feedback loop that sustains good outcomes. A well-planned incremental change creates local benefits that, with effective communication, give the leadership confidence that it should support additional change via additional well-planned increments. Local successes yield progressively broader success that ultimately carries success enterprisewide.

[7] For more details, see Laura H. Baldwin et al., *Incentives to Undertake Sourcing Studies in the Air Force*, Santa Monica, Calif.: RAND Corporation, DB-240-AF, 1998.

[8] For more details, see Brian Stecher et al., *Toward a Culture of Consequences: Performance-Based Accountability Systems for Public Services*, Santa Monica, Calif.: RAND Corporation, MG-1019-EDU, 2010.

making investments to benefit another community in the future. The senior leadership needs to create and sustain a coalition of functional groups that can sustain this effort in a way that ensures that everyone is working toward a common purpose.

The Air Force must pursue this mission in coordination with relevant contractors, which will be pursuing competing goals on many issues. Each Air Force community will engage contractors that mirror the Air Force community. This means contractors will find themselves pursuing parallel efforts within their own channels. The Air Force communities are likely to be successful dealing with their contractor counterparts only if the Air Force can sustain a comparable commitment to invest over the long term. If the senior leadership cannot sustain such an effort, it should rethink the wisdom of seeking more complete technical data and data rights.

One way to test this resolve is to ask whether the senior leadership will tolerate the following kinds of decisions:

- Accept OMB scoring of current expenditures against current budget constraints, even if benefits are not expected for many years.
- Where appropriate, push back on contractor efforts to expand their rights to IP, even if doing so leads to delays in programs and increases in administrative costs induced by contractor efforts to slow roll Air Force efforts. Air Force personnel should do this when contractors assert their rights to data, fail to deliver the full data and rights defined in CDRLs, or mismark technical data in a way that could confuse future users of Air Force rights.
- While doing this, organize data and clarifications of rights in well-organized and managed archives that future users can access to clarify what the Air Force has paid for in the past.
- Where appropriate, be aggressive enough to provoke contractors to take issues to court to test Air Force rights and clarify the legal basis of the Air Force's claims to technical and data rights. Make greater use of Air Force lawyers who understand IP law to clarify where such aggressive tactics are likely to be most helpful.
- Create and sustain cadres of IP lawyers and other experts on technical data and data rights who can fill in details on the tactics associated with the points above and support Air Force personnel with a responsibility to implement these tactics.
- Include audits of activities to protect Air Force rights to data, and ensure delivery of all data specified in CDRLs as part of routine Inspector General (IG) inspections.

To sustain efforts of this kind, the senior leadership may have to maintain closer focus on technical data and data rights over the longer term. One way to do this is to reorganize to separate staffs responsible for short- and long-term decisions. Then allocate resources and responsibilities to their organizations at a higher level, where senior leaders can more easily maintain oversight. For example, the senior leadership could ensure that, each year, the planning, programming, budgeting, and execution process sets aside resources for activities that yield benefits over longer periods and restrict the ability of lower-level personnel to trade these activities away for shorter-term benefits. For example, such an office could be given the rights, responsibilities, and resources to participate in requirements determination, acquisition strategy, contract formation, EMD, and O&S as an advocate for technical data and data rights. The office would police decisions to change agreements about technical data and data rights documented in CDRLs. It would staff teams in regular IG inspections with responsibility for ensuring that Air Force activities are executing policy about data and rights in accordance with higher-level guidance. It could be the home for a cadre of experts on technical data and data rights.

## Change Management Across Organizations and Policy Venues

To succeed in adopting evolved strategies for the acquisition and management of IP, the Air Force leadership must direct change across a broad array of Air Force organizations and policy venues. Leadership investment—taking the form of leadership attention, direction, and resources—must focus on multiple communities to achieve Air Force–wide improvements in performance and cost in the far future. This section describes the most significant organizational locations and policy venues for such change, based on prior RAND research, as well as analysis of the IP strategy options suggested by the Section 809, 813, and 875 Panels or studies and those developed by this report as well.[9]

### *Leadership*

Most significantly, Air Force leadership must decide where and in what amounts to invest dollars in IP based on an assessment of the return on investment from those expenditures. In essence, the Air Force must decide whether it wants to purchase the potential long-term flexibility with respect to operations, sustainment, modifications, and other requirements. This will require weighing these potential long-term benefits against the near-certain increase in pricing for major weapon systems that will accompany more expansive acquisition of IP.

Further, to the extent that the Air Force acquires IP rights, Air Force leadership must ensure that subordinate commands acquire those rights in practice and compel delivery of such rights where necessary and appropriate. Certain rights, such as unlimited rights in OMIT data and FFF data, vest in the government as a matter of law. Others are acquired by operation of DFARS clauses and determinations of funding, as well as the resolution of markings disputes based on assertions by contractors and challenges to those markings by the government. All of these interactions shape the rights that the Air Force will have years later to make use of certain IP and the extent to which the Air Force will have possession of such data and in the right form to utilize. Air Force program offices face real tensions between asserting the government's rights and conceding rights to contractors in consideration for other important facets of contractual performance, such as maintaining schedule progress or cost discipline. However, if the Air Force leadership decides to invest in IP rights to enable greater long-term flexibility with respect to

---

[9] Camm et al., 2021.

operations, sustainment, upgrades, and so on, the Air Force leadership must be willing to emphasize this position throughout the service and ensure that commands, centers, and program offices all act accordingly, with accountability to Air Force leadership for failure to properly acquire and take delivery of necessary IP accompanying major weapon systems.

## Acquisition Policy and Regulations

In addition to command emphasis for the acquisition of IP, the recommendations described in this report may require changes to DoD or Air Force acquisition policy and regulations or interpretations thereof. In recent years, Congress has added a statutory requirement to consider long-term sustainment requirements as part of the IP strategy for major weapon systems.[10] Similarly, in 2018, Congress legislated a requirement for DoD to reach a price agreement with contractors for long-term access to technical data that would facilitate sustainment, at the time of EMD contract award.[11] And Congress added a section to the data rights statute in Title 10 that gives the government an option to purchase technical data needed for "reprocurement, sustainment, modification, or upgrade" (among other purposes) for up to six years after the last article is delivered and "compensate the contractor only for reasonable costs incurred for having converted and delivered the data in the required form."[12] The form of these policy changes matters to the extent that they have a permanence beyond that found in Air Force guidance or directives or even special contract clauses. The same is true of Federal Acquisition Regulation (FAR) and DFARS clauses—there is value in the permanence, transparency, and signaling of using formal regulations to set the terms of the relationship between the Air Force and industry with respect to IP rights.

In this domain, the most significant acquisition policy and regulations are those found in DFARS, specifically clauses 252.227-7013, 252.227-7014, and the related DFARS clauses in Subpart 252.227 relating to licenses and markings. These clauses, along with the prescriptive provisions of DFARS Part 227, operate as both defense acquisition policy and enforceable contractual terms. They govern the relationship between the Air Force and its OEMs. Through flow-down requirements and customary practice among government contractors, these clauses also set the relationships between OEMs and their subcontractors.

Similarly, DoD and Air Force policies relating to acquisition, such as DoDI 5000.02, JCIDS, or Air Force Instruction 63-101/20-101, have special importance because of their permanence, transparency, and importance as governance documents within the service.[13] These documents

---

[10] 10 U.S.C. 2320(e).

[11] 10 U.S.C. 2439.

[12] 10 U.S.C. 2320(b)(9).

[13] Air Force Instruction 63-101/20-101, *Integrated Life Cycle Management*, Washington, D.C.: U.S. Department of the Air Force, May 9, 2017.

do currently address the requirement for the acquisition of IP. AFI 63-101/20-101 states that "[t]he PM shall assess long term IP rights requirements and corresponding acquisition strategies prior to initiating a RFP to acquire systems, subsystems, or end-items to ensure they provide for rights, access, or delivery of data that the Government requires for systems sustainment and to maintain competition throughout the life cycle."[14] The current AFI goes even further to require Milestone Decision Authority (MDA) review for any IP acquisition decisions that do not provide the necessary IP for organic support and sets forth 20 types of operation, sustainment, and modification activity (see Figure 5.1) that must be considered by the program manager (PM) when crafting the Life Cycle Management Plan (LCMP) and Life Cycle Sustainment Plan

**Figure 5.1. Intellectual Property Uses Specified in AFI 63-101/20-101**

> 4.7.1. The PM ensures the program IP strategy, including the performance work statement or SOW for development, production, deployment, and sustainment (for all applicable phases) includes appropriate IP requirements, access, and necessary deliverables, or options for data and equipment deliverables required to support:
>
> 4.7.1.1. Organic source of repair and/or supply decisions.
>
> 4.7.1.2. Government Core depot maintenance capability requirements.
>
> 4.7.1.3. Expeditionary logistics footprint requirements.
>
> 4.7.1.4. Engineering data requirements needed for such activities as integrity programs, sustaining engineering, reliability management, and configuration management.
>
> 4.7.1.5. TOs.
>
> 4.7.1.6. Re-procurement/modification/upgrade.
>
> 4.7.1.7. Demilitarization/Disposal.
>
> 4.7.1.8. Modular open systems approach (MOSA).
>
> 4.7.1.9. Cybersecurity strategies.
>
> 4.7.1.10. Technology refreshment or enhancement.
>
> 4.7.1.11. Training and training program information.
>
> 4.7.1.12. Spare parts procurement.
>
> 4.7.1.13. Testing and Evaluation.
>
> 4.7.1.14. IMD production.
>
> 4.7.1.15. Contractor Logistics Support.
>
> 4.7.1.16. Supply Chain Management.
>
> 4.7.1.17. Depot Level Reparable and consumables procurement.
>
> 4.7.1.18. Support Equipment procurement and maintenance.
>
> 4.7.1.19. Special Tools/Tooling.
>
> 4.7.1.20. Diminishing Manufacturing Sources & Material Shortages (DMSMS).

SOURCE: Air Force Instruction 63-101/20-101, 2017, p. 42.

---

[14] Air Force Instruction 63-101/20-101, 2017, p. 41.

(LCSP) for a given system.[15] The extract from AFI 63-101/20-101 in Figure 5.1 shows these 20 types of operation wherein IP may be required.

Current policy and regulations provide adequate authority for Air Force leaders and program offices to acquire IP to support the life cycle of a major weapon system. However, these laws and regulations do not put a great deal of emphasis on the acquisition of IP relative to other contractual parameters. This balance may deserve adjustment in light of the increasing importance of IP to the operations, sustainment, upgrade, and modification of major weapon systems, such as the sophisticated F-35 platform. There is organizational value in making these changes to department or service policies and regulations, because of the important role these documents play for governing conduct at subordinate levels of the organization. Changing these policies or regulations is a nontrivial effort that may, in the case of regulations like DFARS, require formal rule-making activity engaging departmental leadership and stakeholders. However, the bureaucratic effort required to change these regulations also confers authority and legitimacy, as well as the ability to legally enforce these regulations (and contractual clauses that relate to them) during future engagements with industry.

*Personnel*

The acquisition and sustainment workforce has long been the target of criticism; nearly every commission focused on acquisition reform in the past three decades has recommended improvements for the acquisition workforce.[16]Although much progress has been made in the realm of IP management, more work remains, for two broad reasons. First, IP matters much more than it did 20 or 40 years ago. Software and technical data now matter much more to the combat capability of many major weapon systems because these sophisticated systems rely on software to function. Consequently, the government's rights to have, use, and understand these data have grown in importance. Second, as these data's importance has grown, so too has the complexity of expertise required to understand it and the complexity of laws and regulations relating to it. Within the broad field of acquisition and procurement, IP expertise is quite narrow,

---

[15] The MDA as defined in DoDD 5000.01 is the designated individual with overall responsibility for a program. The MDA has the authority to approve entry of a program into the next phase of the life-cycle process, certify milestone criteria, and is accountable for cost, schedule, and performance reporting to higher authority, including congressional reporting. The MDA for ACAT ID, IC, IAM, and IAC programs is either the defense acquisition executive or service acquisition executive.

[16] For criticism, see President's Blue Ribbon Commission on Defense Management (also known as the Packard Commission), *A Quest for Excellence: Final Report to the President*, Washington, D.C., 1986; Commission on Army Acquisition and Program Management in Expeditionary Operations (also known as the Gansler Commission), *Urgent Reform Required: Army Expeditionary Contracting*, Washington, D.C.: Secretary of the Army, 2007. For improvements, see Susan M. Gates et al., *Analyses of the Department of Defense Acquisition Workforce: Update to Methods and Results Through FY 2017*, Santa Monica, Calif.: RAND Corporation, RR-2492-OSD, 2018; see also John A. Ausink et al., *Air Force Management of the Defense Acquisition Workforce Development Fund: Opportunities for Improvement*, Santa Monica, Calif.: RAND Corporation, RR-1486-AF, 2016.

even among contracting specialists or procurement lawyers. These two trends put the Air Force in a bind: IP arguably matters more now than ever, but the Air Force lacks the personnel to effectively acquire and manage IP.

Congress directed the establishment of IP "cadres" (see Chapter 7) as a solution to this problem. However the Air Force decides to act with respect to these cadres, there exist other opportunities for the implementation of an IP strategy that will require action in the personnel realm. To develop requirements for IP during the acquisition strategy phase and then execute the acquisition of IP during a major weapon system procurement, the Air Force will need an acquisition workforce with greater expertise across the board in IP matters, not just a cadre. The Air Force will also need to align its personnel incentives to reward officers, enlisted personnel, and civilians for effectively acquiring and managing IP and to hold accountable those personnel who do not do so. Within program offices, Air Force leaders must use personnel selection and incentives to elevate successful IP management to the same level as cost or schedule compliance. And the Air Force must invest in its acquisition workforce's IP expertise by continuing to invest in specialized education and training for personnel assigned to program offices, legal offices, and other parts of the service responsible for acquisition and sustainment.

## Acquisition Strategy

Acquisition statute and DoD policy each require the consideration of IP in the development of an acquisition strategy.[17] This consideration shall include both the short-term needs associated with acquisition and operation of a major weapon system, as well as the long-term needs associated with operations, sustainment, upgrades, and modification over the long life of a weapon system. However, within these strategies, there may be prioritizations and trade-offs made among competing priorities, such as cost versus capability. Because IP is a niche area and because its value to programs is often latent and not observed for decades after acquisition, IP rights may struggle for priority or primacy in acquisition strategies. The resulting acquisition strategies emphasize short-term acquisition goals, such as performance, price, and timely delivery, while de-emphasizing long-term acquisition imperatives, such as the procurement of sufficient data rights to support sustainment years in the future. Even more speculative requirements, such as options to acquire IP that can provide the service with flexibility to change its sustainment model or support future upgrades or modifications, receive even less priority and primacy within acquisition strategies. If the Air Force leadership wants to be more strategic

---

[17] 10 U.S.C. 2320(e) states as follows: "The Secretary of Defense shall require program managers for major weapon systems and subsystems of major weapon systems to assess the long-term technical data needs of such systems and subsystems and establish corresponding acquisition strategies that provide for technical data rights needed to sustain such systems and subsystems over their life cycle. Such strategies may include the development of maintenance capabilities within the Department of Defense or competition for contracts for sustainment of such systems or subsystems."

about the acquisition of IP, it must direct the personnel drafting acquisition strategies to elevate the consideration of long-term IP requirements in the strategies they produce, consistent with current statutory and regulatory direction.

*Solicitation, Source Selection, and Contract Formation*

The elevation of long-term IP requirements within acquisition strategy is vital because the requirements specified in acquisition strategies often become requirements in Air Force solicitations and contracts. The most critical decision point for the long-term acquisition of IP is the moment of formation for the EMD contract, when the Air Force agrees with the OEM on the rights in technical data and software to be furnished under that EMD contract. The contractual clauses shaping these rights—including but not limited to contract line item numbers (CLIN) for data, CDRLs, rights in data clauses, and tethering of payments to delivery of data—determine the IP acquired by the Air Force at the birth of a major weapon system. They also govern the IP rights the Air Force will have for the life of the weapon system, unless the Air Force negotiates with the OEM to acquire additional rights beyond those agreed on in the EMD contract.

The formation moment for the EMD contract also provides the Air Force with an opportunity to reach agreement with the OEM on a price for IP at a moment where the Air Force has the benefit of competitive leverage and when the OEM can adjust its pricing to reflect the new balance of rights sought by the Air Force. Indeed, the FY 2018 NDAA directed DoD to "[negotiate] a price for technical data to be delivered [for] development or production" prior to selection of an EMD contractor for a major weapon system.[18] Industry has expressed concern— through leadership by private counsel and bid protest filings—about expansive acquisition of IP by the Air Force.[19] However, the Air Force may be able to work through this opposition by constructively negotiating with industry for IP at the time of EMD contract award, when the parties have the most flexibility, and each side is incentivized to reach a mutually beneficial agreement. In these negotiations, Air Force leaders must ensure that long-term IP requirements receive attention and primacy alongside other important parameters, such as performance, cost, and schedule. The EMD contract that ultimately results from solicitation, source selection, and contract formation should support both the short-term acquisition and operations needs of the weapon system and its long-term operation and sustainment.

---

[18] Sec. 835, FY 2018 National Defense Authorization Act, Pub. L. 115-91, 131 Stat. 1470 (December 12, 2017). In the FY 2019 NDAA, Congress broadened this requirement to also include the sustainment of major weapon systems in addition to EMD contracts.

[19] W. Jay DeVecchio, "Data Rights Assault: What in the H (Clause) Is Going on Here? Air Force Overreaching on OMIT Data," *Government Contractor*, Vol. 60, No. 2, January 17, 2018, pp. 1–6.

*Engineering and Manufacturing Development Contract Execution and Performance*

One of the most important venues for change management is the execution and performance of the EMD contract. The decisions made during the performance of this contract effectively determine the Air Force's IP rights and possession thereof for decades to come. Recent studies indicate that Air Force program offices are frequently prioritizing performance, cost, and schedule compliance over other contractual parameters, such as complete delivery of IP required under CDRLs. In doing so and continuing to pay contractors for their performance, program offices are effectively waiving the Air Force's rights to technical data, as well as the rights to take delivery of that data. Likewise, when contractors deliver data with proprietary markings or markings indicating restrictions on the government's ability to use data, the Air Force has an obligation under current DFARS provisions to challenge those markings immediately or else forfeit the right to do so. There is evidence in many major programs to suggest that Air Force program offices are not pushing back against contractors and disputing markings where the service has grounds to do so. Worse, there is some evidence to suggest that the Air Force is not preserving information about data rights, data delivery, funding for the creation of data, and markings thereof that would enable subsequent generations of Air Force personnel to successfully assert rights in data or successfully challenge restrictive rights assertions by contractors. And it is unclear whether the Air Force is adequately tethering the delivery of technical data to payments, depriving the service of its primary leverage to compel delivery of IP during performance of the contract or at the time of contract termination. Consequently, the Air Force may be losing a considerable volume of IP through a slow process of neglect, in which rights are not asserted and contractor assertions are not challenged.

*Solicitation, Contract Formation, and Performance of Sustainment Contracts*

Just as the formation moment for EMD contracts presents an opportunity to reach agreement on the allocation of IP rights and the price for such rights, so too do sustainment contracts with the OEM offer such an opportunity. Although the OEM has something of a monopolist position after the EMD contract, the Air Force may be able to leverage the competitive pressure of sustainment contract awards to revisit the service's rights in data and delivery of the same. The Air Force may also be able to use these formation moments for sustainment contracts with OEMs to affirm the allocation of rights from prior contractual rounds or understandings related to OMIT and FFF data or requirements with respect to the delivery of data. This opportunity is vitiated by the OEM's monopoly on IP that exists after the EMD contract is awarded.

*Organizational Metrics and Incentives*

The current acquisition system places its ultimate emphases on cost and schedule, followed closely by combat capability (performance), operational readiness, and other parameters. The adequacy of IP to support decision flexibility on sustainment or to support future service requirements with respect to diagnostics, investigations, upgrades, or modifications does not rise to the same level as these other priorities in the formation of acquisition strategy. However, there is increasing evidence that the sustainment of major weapon systems over their life cycle exceeds the acquisition cost and that the availability of IP to enable flexibility in sustainment plays a key role in determining the cost of sustainment too.[20] Consequently, it is important for Air Force leaders to prioritize the acquisition of IP. As described above with respect to change management, Air Force leaders must deliberately incentivize acquisition and procurement offices to make this change. Such incentives should include but not be limited to appropriate inclusion of IP acquisition as a factor in personnel evaluations; consideration of performance with respect to IP acquisition in selections and promotions; and recognition of exceptional performance by different parts of the Air Force with respect to IP acquisition, in order to stimulate intraservice competition. Alongside these incentives, Air Force leaders should consider metrics for performance, such as senior leadership visibility on IP acquisition in reporting regarding major weapon systems; inclusion of IP acquisition requirements as part of command inspection lists and IG inspection lists for regular audits and inspections; and accountability systems such as after-action reviews and published case studies for instances of good and bad IP acquisition.

---

[20] Light et al., 2018. Camm et al., 2021.

# 6. A Congressionally Mandated Solution: Intellectual Property Cadres

In Section 802 of the FY 2018 NDAA, Congress directed the Secretary of Defense to create a "cadre of intellectual property experts" to help the department better manage data rights issues.[1] This statutory mandate states that "[t]he purpose of the cadre is to ensure a consistent, strategic, and highly knowledgeable approach to acquiring or licensing intellectual property by providing expert advice, assistance, and resources to the acquisition workforce on intellectual property matters, including acquiring or licensing intellectual property." The statute sets forth several duties for the cadre to perform as part of the acquisition system for the department, mandating that the cadres shall

(A) interpret and provide counsel on laws, regulations, and policies relating to intellectual property;

(B) advise and assist in the development of an acquisition strategy, product support strategy, and intellectual property strategy for a system;

(C) conduct or assist with financial analysis and valuation of intellectual property;

(D) assist in the drafting of a solicitation, contract, or other transaction;

(E) interact with or assist in interactions with contractors, including communications and negotiations with contractors on solicitations and awards; and

(F) conduct or assist with mediation if technical data delivered pursuant to a contract is incomplete or does not comply with the terms of agreements.[2]

The statute creates some tension between the Office of the Secretary of Defense (OSD) and the services as to who should oversee the cadres on a day-to-day basis. In one part, Section 802 states that the Under Secretary of Defense for Acquisition and Sustainment "shall establish an appropriate leadership structure and office within which the cadre shall be managed and shall determine the appropriate official to whom members of the cadre shall report." But in the next clause, Section 802 states that "[t]he cadre of experts shall be assigned to a program office or an acquisition command within a military department." The remaining parts of Section 802 empower OSD to ensure the technical competence of the cadres, provide details as necessary, use DoD special hiring authorities to support the cadres, and enter into specialized contracts

---

[1] See Section 802, FY 2018 NDAA; see also U.S. House of Representatives, National Defense Authorization Act for Fiscal Year 2018 Conference Report to Accompany H.R. 2810, Washington, D.C., November 2017, p. 168.

[2] Section 802, FY 2018 NDAA.

(such as with external consultants or private counsel) to support the cadres.[3] Further, Section 802 authorizes OSD to use the Defense Acquisition Workforce Development Fund for the purposes of recruiting, training, and retention of the cadre and for the compensation of cadre personnel for up to three years.

In the conference report that accompanied the FY 2018 NDAA, Congress indicated a desire to see this provision result in a new "Office of Intellectual Property within the Department of Defense to standardize the Department's approach toward obtaining technical data, promulgate policy on IP, oversee the cadre of IP experts, and serve as a single point of contact for industry on IP matters."[4] The conference report further stated that the new cadre provision "would add IP positions to the acquisition workforce" and revise training provided to the acquisition workforce on IP matters. The provision originated in the House and was not mirrored in the Senate by similar language. In reconciling their bills, conferees agreed that OSD would "establish an appropriate organizational structure to support the cadre of intellectual property experts," although this did not resolve how the structure created by OSD would then reside within the services at the program offices or command level. Congress added that it wanted to see OSD use the Defense Acquisition Workforce Development Fund "to expand access to training and educational opportunities" relating to IP for the acquisition workforce. And, notably, Congress signaled that it expected OSD to designate a central point of contact within DoD to engage with industry on IP matters and a congressional expectation that "shall regularly engage with appropriately representative entities, including large and small businesses, traditional and nontraditional Government contractors, prime contractors and subcontractors, and maintenance repair organizations" on IP issues of concern to these entities.[5]

Beyond this broad vision for IP cadres within the department and some guidance on how DoD should pay for their creation, Congress did not specify the precise organizational parameters for their construction. More importantly, Congress did not mandate precisely how the services should integrate these cadres into their day-to-day operations, including incentives for their use by supported offices and how to evaluate their efficacy in supporting service and DoD goals. This chapter examines a few existing cadre models for the purposes of understanding how

---

[3] On technical competence: "[T]he Under Secretary shall ensure the cadre has the appropriate number of staff and such staff possesses the necessary skills, knowledge, and experience to carry out the duties under paragraph (2), including in relevant areas of law, contracting, acquisition, logistics, engineering, financial analysis, and valuation."

On hiring authorities: "The Under Secretary may use the authorities for highly qualified experts under section 9903 of title 5, to hire experts as members of the cadre who are skilled professionals in intellectual property and related matters."

On external consultants: "The Under Secretary may enter into a contract with a private-sector entity for specialized expertise to support the cadre. Such entity may be considered a covered Government support contractor, as defined in section 2320 of this title."

[4] U.S. House of Representatives, 2017, p. 1909.

[5] U.S. House of Representatives, 2017, p. 1910.

similar cadres have worked in other contexts and distills some potential recommendations for the Air Force as it considers how best to structure its IP cadre to respond to Section 802 of the FY 2018 NDAA.

This report defines a cadre broadly to include a group of subject-matter experts who can be drawn on individually or as a group to provide their knowledge, skills, or abilities to their organization (or another) when needed. Cadre personnel typically have knowledge, skills, or abilities that are both greater and more specialized than the typical "general purpose" person working within their organization. This report examines a few cadres as they have been used by the department, including the Defense Contract Audit Agency (DCAA), the Navy Price Fighters, the Naval Air Command (NAVAIR) Competency Aligned Organizations (CAOs), and FFRDCs. The overview of each cadre will discuss the organization and function of each cadre as well as the model's applicability to the issue of data rights.

### Defense Contract Audit Agency

DCAA was started in 1965 by Secretary of Defense Robert McNamara in order to streamline the defense contract auditing process across DoD.[6] As the purpose of the agency is to centralize resources for a niche activity, this organization falls under the working definition of a cadre presented in the previous section. As a cadre, DCAA has an organization with a highly centralized function but a highly decentralized organizational structure.

DCAA is responsible for auditing all defense contracts for all DoD service branches and the Defense Contract Management Agency (DCMA).[7] The agency effectively centralizes the specialized task of contract auditing across DoD into one agency, helping to decrease redundancies and oversights. While DCAA has a relatively broad reach in terms of the groups it supports, it has a relatively narrow set of functions. DCAA only audits contracts. It does not complete internal compliance reviews.[8] This narrow, specialized function indicates that, using the definition provided at the beginning of the chapter, DCAA is a cadre.

While DCAA serves to centralize contract auditing across DoD, the organizational structure of DCAA is decentralized to better allocate resources. DCAA has created a system in which auditors are organized by both region and contractor size. The agency has three regional hubs, which oversee suboffices located in the geographical area, and four Corporate Audit Directorates, which manage the contracts associated with the largest defense contractors. These two organizational divisions have some overlap as offices may be established at a contractor's site to

---

[6] Defense Contract Audit Agency, "About DCAA," webpage, undated-a.

[7] Defense Contract Audit Agency, *Fiscal Year (FY) 2019 President's Budget: Operation and Maintenance, Defense-Wide,* Fort Belvoir, Va., February 2018a, p. DCAA-57.

[8] Defense Contract Audit Agency, 2018a, p. DCAA-57.

focus exclusively on that contractor's contracts.[9] The decentralized structure of DCAA allows for the better allocation of resources to account for the large number of audits it must complete for a large array of contractors. Importantly for the purposes of this analysis, DCAA is a defense agency that supports the services but does not reside within any of the services.

Staff are an important asset for DCAA as they are the primary resource needed for the completion of contract audits. As a cadre, the agency must retain skilled employees to quickly and efficiently respond to audit requests. As a result, the majority of the expenses associated with running this cadre are tied to the cost of labor.[10] Staff must have in-depth knowledge about contract auditing to effectively complete their jobs. Therefore, auditing staff are required to have a bachelor's degree and at least four years of experience in accounting or auditing.[11] These qualifications allow the agency to maintain the proper level of specialization and experience necessary for it to qualify as a cadre.[12]

DCAA does not charge DoD elements or contractors for its services. Instead, its funding is provided through the Defense-Wide President's Budget.[13] Audits therefore do not create any additional direct costs for a program nor direct costs for a service when DCAA assets are requested (directly or indirectly). For those DCAA activities that are mandated by law, regulation, or contract, DCAA initiates its activities based on regulatory or contractual requirements (such as those for an incurred cost submission or rate review) and is thus considered mandatory. Programs and services have the option though to request additional DCAA support as may be needed to help with cost or pricing analysis or auditing.

## Navy Price Fighters

The Navy Price Fighters were formed in 1983 to push back against the perceived high prices of spare parts and create a "should cost" pricing system for spares. Since its foundation, the organization has shifted its focus from analyzing spare prices to analyzing prices for any type of acquisition.[14] The Price Fighters function as a cadre due to their specialized function and staff.

---

[9] Defense Contract Audit Agency, *Report to Congress on FY 2017 Activities Defense Contract Audit Agency*, Fort Belvoir, Va., March 31, 2018b, p. 2.

[10] Defense Contract Audit Agency, 2018a, p. DCAA-74.

[11] Defense Contract Audit Agency, "Qualifications," webpage, undated-b.

[12] DCMA offers a similar case study as DCAA, because of the way DCMA is frequently brought in to do the contracting and other associated functions for Major Defense Acquisition Programs. See DCMA Manual 3101-01, *Program Support Life Cycle*, Washington, D.C.: U.S. Department of Defense, October 23, 2017, change 1, September 20, 2018; DCMA Instruction 3101, *Program Support*, Washington, D.C.: U.S. Department of Defense, July 28, 2017, change 1, September 20, 2018.

[13] Defense Contract Audit Agency, 2018a, p. DCAA-57.

[14] Naval Supply Systems Command Weapon Systems Support, "Navy Price Fighters," Norfolk, Va.: Naval Supply Systems Command Weapon Systems Support, February 2018, slide 3.

The Price Fighters are a small organization exclusively based out of Norfolk, Virginia, that employs around 50 to 60 personnel.[15] The small size of the organization emphasizes the narrow mission focus of the group and also allows for the staff specialization needed to complete this function. The narrow, centralized structure of the Price Fighters provides the organizational focus necessary for a cadre, despite the wide variety of programs supported.

The Price Fighters emphasize the technical background of personnel as staff need to have in-depth knowledge of manufacturing processes to know the should-cost price of a part. This means that most of the staff have substantial experience working in defense manufacturing and usually have a particular trade or specialty, such as radar or electronics, that meshes with their current Price Fighter assignment.[16] Since the Price Fighters support a wide variety of programs, the skills-based organization of staff creates a logical system of resource management.

Price Fighters are brought in when a program believes that the prices charged by a supplier for a part are too high. The ad hoc nature of the Price Fighters' work indicates that the costs of the group are neither budgeted into the cost of a program nor supplied funding through the NDAA. Instead, the cost of the Price Fighters is an additional cost for a program, billed by total hours worked.[17] However, as the Price Fighters' work may lower the overall cost of a project, the cost of the work completed by the group may be offset by program savings.

As the Price Fighters are organized under the Naval Supply Systems Command, the majority of their work supports naval programs. However, the Price Fighters may work with any government agency, including civilian agencies, as long as the work completed is consistent with the Price Fighters' mission.[18]

## Naval Air Systems Command Competency Aligned Organizations

NAVAIR was established in 1966 under Secretary of Defense Robert McNamara to manage naval aviation acquisition and sustainment programs. The organization continues to manage the full life cycle of naval aviation systems. In the 1990s, NAVAIR reorganized itself to overcome perceived shortfalls associated with the traditional, centralized "stovepipe" organization present in other defense acquisition organizations. Instead, NAVAIR organized itself into (1) CAOs, which are decentralized teams that focus on specific skills, and (2) Integrated Program Teams

---

[15] Director, Acquisition Workforce Management, "ASN(RDA)PCD 1102s Rotations at Price Fighters," Assistant Secretary of the Navy for Research, Development, and Acquisition, Washington, D.C., undated.

[16] Michael Boito, Kevin Brancato, John C. Graser, and Cynthia R. Cook, *The Air Force's Experience with Should-Cost Reviews and Options for Enhancing Its Capability to Conduct Them*, Santa Monica, Calif.: RAND Corporation, TR-1184, 2012, p. 20.

[17] Boito et al., 2012, p. 20.

[18] Boito et al., 2012, p. 32.

(IPTs), groups that support a specific program.[19] While program-specific responsibilities are focused within the IPTs, certain aspects related to a program are outsourced or supported by the CAO. In general, these aspects support the core functions of the IPT but may require a more specialized background or set of knowledge.[20] Therefore, as the CAOs serve to centralize access to a specialized resource, they can be thought of as cadres functioning within the larger NAVAIR organization.

Currently, NAVAIR has eight separate CAOs, each of which provides support and resources to the larger organization. These CAOs are program management, contracts, research and engineering, test and evaluation, logistics and industrial operations, corporate operations, comptroller, and office of counsel.[21] Each CAO covers a potential problem area for a program that may require more specialized knowledge or resources than the staff of that program possess. These competencies serve to centralize resources across the organization, allowing for better overall resource allocation across programs.

Each competency provides several different services to the greater organization. First, CAOs may provide specialized services to a particular program staff. Second, CAOs may provide access to specialized facilities, such as testing facilities. Third, CAOs may provide training on the issues in which the CAO specializes.[22] The resources and services provided by the CAO help to prevent information silos and redundancies within the organization by streamlining resource access across the organization and ensuring that different programs within NAVAIR have access to the same information and resources.

NAVAIR's main office is located in Patuxent, Maryland, with nine supporting facilities located across the United States and abroad.[23] This model presents a midpoint in terms of locational organization between the extreme decentralization of DCAA and the extreme centralization of the Navy Price Fighters. As mentioned in the previous paragraph, NAVAIR organizes personnel by skill set and program office. Therefore, having a relatively centralized system of facilities allows for better cross-utilization of skilled personnel by the different program offices.

---

[19] Naval Air Systems Command, "Overview," Patuxent, Md.: U.S. Navy, undated; Beth Springsteen and Elizabeth K. Bailey, *The F/A-18E/F: An Integrated Product Team (IPT) Case Study*, Alexandria, Va.: Institute for Defense Analysis, NS D-8027, April 1998, pp. 11–12.

[20] Naval Air Systems Command, "1.5 Role of the Competency Aligned Organization," in *NAVAIR—Integrated Program Team Manual: Guidance for Program Teams and Their Subsets December 1996*, Patuxent, Md.: U.S. Navy, April 10, 2014.

[21] Naval Air Systems Command, *NAVAIR Acquisition Guide 2014/2015*, Patuxent, Md.: U.S. Navy, October 1, 2013, p. 4.

[22] Naval Air Systems Command, 2013, p. 4.

[23] Naval Air Systems Command, undated.

NAVAIR is funded through the appropriation of funds from the Special Program Office.[24] Therefore, use of NAVAIR resources can be thought of as part of the program's overhead. While NAVAIR primarily works the Navy and Marines, it may accept interdepartmental funding to work on programs that are cosponsored by multiple military branches.[25] This allows more flexibility in terms of both funding and mission.

## Federally Funded Research and Development Centers

FFRDCs are federally funded centers that focus on research and development (R&D) issues outside a federal agency's scope of work. FFRDCs have existed since World War II.[26] The specialized nature of FFRDC work and the centralization of talent that is associated with the FFRDC means that they are cadres using the definition provided at the beginning of the chapter. The three different types of FFRDC are R&D laboratories, study and analysis centers, and systems engineering and integration centers.[27] Even though these three different FFRDC types have different organizational structures and purposes, they can all be thought of as cadres because they all serve as specialized centers that complete work for the government.

FFRDCs must be sponsored by an agency within the government. As a result, the mission and funding associated with the FFRDC is tied to the sponsoring agency's mission and funding status. Furthermore, the agency-specific mission and funding ties lead to the promotion of long-term contracts and relations between FFRDCs and agencies.[28] These long-standing agency ties indicate that the FFRDC has the necessary functional focus to be categorized as a cadre.

The mission-specific focus and emphasis on long-term relationships also affects the staffing requirements for the FFRDC. In particular, the "stability and continuity" that the long-standing relationships between the FFRDC and the sponsoring agency promote allow for the FFRDC to "attract high-quality personnel" and "maintain in-depth expertise" on the issues that the FFRDC is tasked to address.[29] Due to the variety of tasks and purposes an FFRDC may be presented with, personnel have specialized yet varied backgrounds.

Another facet of the FFRDC that emphasizes its position as a cadre is the fact that these organizations are limited in terms of funding and responsibility in order to hone the functional

---

[24] Naval Air Warfare Center, "Funding Documents," Orlando, Fla.: U.S. Navy, Training System Division, August 15, 2013.

[25] Naval Air Warfare Center, 2013.

[26] Marcy E. Gallo, *Federally Funded Research and Development Centers (FFRDCs): Background and Issues for Congress*, Washington, D.C.: Congressional Research Service, R44629, December 1, 2017, p.1.

[27] Gallo, 2017, pp. 2–3.

[28] U.S. General Accounting Office, *Federally Funded R&D Centers: Information on the Size and Scope of DOD-Sponsored Centers*, Washington, D.C.: U.S. General Accounting Office, National Security and International Affairs Division, B-270464, April 1996.

[29] Gallo, 2017, p. 3.

focus necessary for the proper maintenance of the FFRDC's mission and intent. This is to prevent "mission creep," which occurs when the federal government spends too much money on FFRDCs and leads to the degradation of the private sector's ability to innovate and provide solutions to the government.[30] As a result, the amount of funding for FFRDCs may be limited by Congress. For example, Congress has limited the total number and the amount of funding available to DoD FFRDCs since 1993.[31] While this funding and organization limitation preserves the deep focus necessary for a cadre, it also means that amending FFRDC functions to complete a new, specialized mission is difficult as any additional task would necessarily involve an equal de-obligation of resources from a preexisting task.

The primary agencies that sponsor FFRDCs are DoD and the Department of Energy. However, there is no limit on which agencies may sponsor FFRDCs, and several other agencies, such as the Department of Homeland Security and the National Science Foundation, sponsor FFRDCs.[32] Furthermore, FFRDCs may perform work for agencies other than the sponsor. However, the extent of this work tends to be limited as the FFRDC's primary function is to support its sponsor. Funding for work performed for other agencies is provided by the agency in question rather than the sponsoring agency.[33] These funding requirements emphasize the support and mission-specific role of the FFRDC.

## Relevance to Intellectual Property Issues

While the cadre formats outlined above represent valid structures for organizing and managing a cadre, not all of the tenets of these cadres may be relevant to the issue of data rights. Furthermore, each cadre type is faced with challenges specific to its format and function.

An assumption within these models is that the service can effectively recruit, retain, and manage the specialty personnel needed to staff an IP cadre. Historically, the services have struggled to recruit, retain, and effectively manage their acquisition workforces, particularly those high-skill persons with expertise in great demand by the private sector.[34] To recruit and

---

[30] Gallo, 2017, p. 3.

[31] Gallo, 2017, p. 3.

[32] Gallo, 2017, p. 2.

[33] Gallo, 2017, p. 12.

[34] For a discussion of these issues, see Susan M. Gates et al., *The Defense Acquisition Workforce: An Analysis of Personnel Trends Relevant to Policy, 1993–2006*, Santa Monica, Calif.: RAND Corporation, TR-572-OSD, 2008; Susan M. Gates et al., *The Department of the Navy's Civilian Acquisition Workforce: An Analysis of Recent Trends*, Santa Monica, Calif.: RAND Corporation, TR-555-NAVY, 2009; Susan M. Gates et al., *Analyses of the Department of Defense Acquisition Workforce: Update to Methods and Results through FY 2011*, Santa Monica, Calif.: RAND Corporation, RR-110-OSD, 2013; Christopher Guo, Philip Hall-Partyka, and Susan M. Gates, *Retention and Promotion of High-Quality Civil Service Workers in the Department of Defense Acquisition Workforce*, Santa Monica, Calif.: RAND Corporation, RR-748-OSD 2014; Gates et al., 2018.

retain contracting officers, engineers, and lawyers with IP expertise, the Air Force must compete effectively against industry, academia, and the private legal sector and do so with monetary and nonmonetary incentives that may be inadequate next to what the private sector can offer.

Notwithstanding the challenges of recruiting and retaining cadre personnel, in the realm of cadre performance, challenges generally relate to either the availability of resources or the scope of work. These issues are somewhat correlated as a lack of resources may affect a cadre's ability to complete its mission or an overly broad mission may place strain on the cadre's resources. The following sections will analyze how resource and scope problems may affect a cadre's ability to function within the setting of IP.

### Staffing and Employment

The need for support from an IP cadre is likely to vary over the life cycle of a program. The cadre will provide varying forms of direct support in each phase of the program. It will also provide active training to Air Force personnel for their IP-related roles in each phase of a program. Relatively modest support during requirements determination is likely to ramp up during the formation of an acquisition strategy and IP strategy derived from it. It will remain active during market research and planning for a source selection, peaking during the evaluation and negotiation of final terms of offers. Vigilance will be required during the execution of EMD to protect data and data rights acquired earlier to oversee delivery of data. If arrangements have been properly structured in earlier phases, demand should recede during O&S.

A second cycle will also be important to the cadre at a different level of the Air Force. That is the planning, programming, budgeting, and execution system cycle, during which the cadre should remain active in providing action officers who can advocate for sustaining the resource the Air Force needs to invest in IP repeatedly through the lives of each of its programs. Some IP cadre personnel will likely become experts in planning, programming, budgeting, and execution system processes and use that expertise to maintain tight ties between an enterprisewide focus on IP issues and the instantiation of that focus in individual programs, particularly earlier in their life cycles.

These activities will not generate steady state demands on the IP cadre. So a large, decentralized, on-call staff like that which DCAA uses to organize its labor resources both regionally and around major defense contractors would not work well.[35] A large standing staff requires a steadier workload to assure staff retention. A smaller, more centralized workforce would most likely provide better staffing and retention results for the IP cadre. This model is similar to the staffing model used by the Navy Price Fighters, which relies on a small, highly

---

[35] Defense Contract Audit Agency, 2018b, p. 2.

specialized staff to handle periodic problems associated with part costs.[36] IP issues are a niche issue that tends to be focused around the beginning portion of a program's life cycle. Therefore, a staffing model that provides for a smaller, more specialized cadre will most likely better fit the needs of the data rights cadre.

### Funding

Funding is another major issue that affects the ability of a cadre to effectively function and utilize resources. The cadre models reviewed in the previous section provide several different funding models that may or may not work well for an IP cadre. These models can be roughly divided into models that provide either budgeted funding for the cadre or ad hoc funding for the cadre based on work completed.

In general, ad hoc funding works best for cadre models that complete activities that are not guaranteed to arise during a program's life. This can be seen in the Navy Price Fighters model, which provides assistance to a program when the program faces problems with the cost of a part.[37] As there is no guarantee that the Price Fighters will be called in during a program's life cycle, their operations costs are not accounted for in a program's budget or through congressional appropriations. However, part of the purpose in establishing an IP cadre is to standardize the treatment of IP issues across programs rather than to address these issues only in instances when a contractor is not cooperating. Therefore, ad hoc funding for the IP cadre may not be the proper choice.

Cadres that provide services central to the acquisition or life cycle of a program tend to have their funding provided as a portion of the total project budget. For example, NAVAIR, which provides support to a project's program executive officer, has its funding provided from the budget of relevant programs.[38] As reviewing IP is an integral part of a major weapon system's acquisition process, the cadre provides a service that is necessary for the proper completion of the acquisition process. Therefore, program funding should most likely be provided from the budget of the program under question. However, the language of Section 802 also provides the ability for DoD and the Air Force to centrally fund these from the Defense Acquisition Workforce Development Fund for the start-up phase of the cadre(s).

### Cadre Mission and Scope

Cadres vary greatly in terms of the specificity of their mission. One of the main aims of the IP cadre is to provide a "consistent, strategic, and highly knowledgeable approach" to program

---

[36] Boito et al., 2012, p. 32.

[37] Boito et al., 2012, p. 20.

[38] Naval Air Systems Command, undated.

IP issues.[39] Therefore, a properly outlined mission is integral to the proper functioning of the IP cadre. The IP cadre will need to decide how broad or narrow its scope of work will be in order to function properly and consistently.

Such cadres as FFRDCs and NAVAIR have relatively broad missions that allow for a larger degree of flexibility in terms of work that may be completed. An advantage of this flexibility is that these types of cadres may be able to address issues that are aligned with their overall goals but not specifically addressed by their mission. However, a broad mission may lead to a cadre losing focus on its primary objectives, a phenomenon known as "mission creep," which leads to a cadre infringing on the mission of other, supporting organizations.[40] In the case of an IP cadre, an overly broad mission may lead to the IP cadre infringing on the ability of the system program office or other acquisition cadres, such as DCAA, to properly complete their jobs.

By nature, an IP cadre has a somewhat narrow mission focus as it must focus on issues related to IP. However, it is unclear how specific the cadre's focus should be. In general, the cadre's mission should be specific enough to provide a clear, consistent outline of the work that will be completed by the cadre but broad enough to encompass special cases that align with its overall goals.

## Summary

Table 6.1 summarizes basic characteristics of four different types of cadres currently employed in DoD. They offer a context in which to consider what a new IP cadre might look like in the Air Force. The new IP cadre will face a variety of issues related to both its use of resources and its overall mission and scope of work. According to the research on cadres completed for this report, the Air Force IP cadre should have the following characteristics:

- a specialized, centrally located staff that can execute their duties at individual work sites (i.e., program offices) as necessary or appropriate
- funding provided from the defense acquisition fund during the start-up phase and then funding provided by budgets of relevant programs
- a narrow mission statement that provides a consistent definition of the work to be completed by the cadre
- support from Air Force leadership to motivate the cadre's effective use by supported commands and agencies across the service.

---

[39] U.S. House of Representatives, 2017, p. 168.

[40] Gallo, 2017, p. 12.

## Table 6.1. Cadre Summary

| Issue | DCAA | Price Fighters | NAVAIR CAOs | FFRDCs |
|---|---|---|---|---|
| Personnel | College educated with auditing experience | Skilled tradesmen with technical experience | Experts in IPT-specific field (e.g., engineering, management) | Researchers with background in relevant discipline |
| Facility organization | Highly decentralized; organized by region and contractor | Highly centralized; one location in Norfolk, Va. | Centralized; central office with branch offices | Varies by FFRDC |
| Mission | Specific; complete contractor audits | Specific; find should-cost value for items | Broad; support program executive officer acquisition efforts | Broad; support sponsoring agency's R&D needs |
| Funding | Provided by Congress through the President's budget | Charge hourly for services | Part of program costs | Percentage of sponsoring agency's funding; congressionally capped |
| Sponsoring agency | All DoD or defense acquisitions are eligible | Primarily Navy/Marines; may work on other DoD or federal projects | Primarily Navy/Marines; may work on projects cosponsored by another military service | Primarily DoD and Department of Energy; may work on other federal projects with limitations |

# 7. Long-Term Option Pricing

This chapter discusses the viability of using options contracts to mitigate the Air Force's technical data challenges. As discussed in Chapter 5, the Air Force is in the best negotiating position to purchase technical data packages and associate rights during EMD source selection, when competition between offerors drives down prices. And in the FY 2018 NDAA, Congress directed DoD to "[negotiate] a price for technical data to be delivered [for] development or production" prior to selection of an EMD contractor for a major weapon system.[1] However, the Air Force has limited information regarding what technical data it needs because such needs do not arise until years or decades after source selection.[2] Since the 1990s, the Air Force (and other armed services) have generally planned for OEM support of new weapon systems over their lifetimes without properly considering uncertainty about the quality and cost of that support. This has led to systematic underinvestment in technical data during the source selection period. While this preference avoided purchasing technical data and rights that would never become useful, it also led to vendor lock that has been hard to reverse when unanticipated technical data needs arose.

Options contracts potentially obviate the need for precautionary underinvestment by granting the Air Force the right to purchase technical data and rights at competitive prices at the time a need for such data arises. But such rights may themselves be costly, and the Air Force will need to negotiate several obstacles to take full advantage of options contracts.[3]

## What Are Options Contracts?

An options contract is an agreement between a purchaser and a seller that gives the purchaser the right to buy a particular asset in the future at a price that the parties agree on ahead of time. Options contracts are important tools in transactions involving future uncertainties, most notably in the securities, commodities, and real estate markets.

---

[1] Sec. 835, FY 2018 National Defense Authorization Act, Pub. L. 115-91, 131 Stat. 1470 (December 12, 2017). In the FY 2019 NDAA, Congress broadened this requirement to also include the sustainment of major weapons systems in addition to EMD contracts.

[2] See supra at Chapter 2.

[3] The Section 813 Panel Report addresses options contracts and related alternatives in Tension Point Papers 24, "Deferred Ordering Period: 6 Years (Rather Than Perpetual)" and "Deferred Ordering Part 1: Only Data 'Generated' Under the Contract"; 25, "Time Limits on Priced Contract Options"; 26, "Deferred Ordering Part 2: All Interface or Major Systems Interface Data May Be Ordered Regardless of U.S. Government Development"; 27, "Failure to Define and Order CDRLs (Reliance on Deferred Ordering and DAL to Obtain Data)"; and 28, "Escrow as a Form of Deferred Delivery."

Options contracts are valuable because they enable a purchaser to hedge against future uncertainty. For example, a real estate developer may be interested in purchasing land and has several plots to choose from. But each plot is useful only if the developer can obtain the requisite city permit for that plot. If he purchases a plot now but fails to obtain a permit, he would have paid a substantial sum for a useless asset. However, if he waits until he obtains a permit for a specific plot before purchasing it, the landowner becomes a monopolist and can raise the asking price or even refuse to sell.

The developer can use options contracts to solve this dilemma: he pays the landowners for the right to purchase land in the future for an agreed on price. The amount he pays for the option is called the *premium*," and the agreed on price for the land is called the *option price*. The developer will always pay the option premium, but he pays the option price only if he exercises the option. The premium and option price are likely to be reasonable because they were set during competition. If the developer is able to obtain the permit, he exercises the option and purchases the plot for a reasonable option price. And if the permit falls through, he has no need for the land and so will decline to make the purchase. For the cost of the option premium, he has insulated himself from uncertainty.

## How Options May Mitigate the Air Force's Intellectual Property Challenge

The Air Force faces a problem that bears similarity to that of the real estate developer: it must decide what technical data packages and associated rights to purchase at EMD source selection. While the Air Force can secure a reasonable price for weapon systems data if there are competing bidders, it cannot accurately predict what technical data and rights will ultimately be needed. This is because many technical data needs do not arise until well into a weapon system's multidecade life cycle.[4] If the Air Force is aggressive in purchasing technical data during source selection, it risks paying for data it will never use because no need ever arises. This is akin to the developer who pays for a plot of land only to find the land is useless because he could not get the permit. And if the Air Force declines to purchase technical data during source selection, by the time the need for such data arises, the Air Force will face a sole-source OEM that can charge exorbitant prices or simply refuse to sell. This second scenario is akin to the developer who faces a monopolist landowner because he waited until after he obtained a specific permit for the landowner's land.

For many years, DoD regulation and policy has effectively required the Air Force to favor the second type of risk: that is, underinvesting in technical data during source selection. DFARS specifically directs Air Force program offices "to acquire only technical data, and the rights in

---

[4] See supra at Chapter 2.

that data, necessary to satisfy agency needs."[5] That message is repeatedly conveyed to program officers in countless directives, instruction, and manuals.[6] Under this standard, program offices are not authorized to purchase technical data that merely have a likelihood of promoting an agency need. Rather, the data must be "necessary to satisfy agency needs."[7] But the activities for which technical data are needed—described in Chapter 2—typically do not occur until decades after source selection. And they may never occur at all. As such, it is virtually impossible for program offices to predict what technical data will become "necessary to satisfy agency needs." Put another way: the breadth of potential activities during a weapon system's life cycle for which technical data are required is far broader than activities that reasonably could be deemed necessary at source selection. Thus, a program office will face pressure to underinvest in technical data.

In response to perceived underinvestment in technical data and licenses thereof, Congress amended the federal statute governing rights in technical data, directing services to program offices to assess long-term technical data needs and establish acquisition strategies that consider technical data requirements for weapon system sustainment over their entire life cycle.[8] Congress further added a provision that requires acquisition strategies to "address the merits of using a priced contract option for future delivery of technical data that were not acquired upon initial contract award."[9]

Priced options contracts enable the Air Force to establish an acquisition strategy that better accounts for long-term technical data needs. As explained above, the current preference for underinvestment in technical data is caused, in part, by the inability for program offices to predict accurately future activities for which technical data may be needed. But options contracts obviate the need for accurate prediction. Rather than purchasing technical data only for future activities that a program office determines to be "necessary," the Air Force may purchase the option to acquire data for activities that are merely *likely* or *foreseeable*. It would not exercise those options unless such data become required. And because the option price for data is negotiated during competition, it is likely to be far more reasonable than the price obtained

---

[5] DFARS 227.7103-1(a): "DoD policy is to acquire only the technical data, and the rights in that data, necessary to satisfy agency needs."

[6] See, for example, U.S. Air Force Space and Missile Systems Command, 2018.

[7] See Valerie Insinna, "Lockheed Loses Out on Its US Air Force Huey Replacement Protest," *Defense News*, May 22, 2018 (describing Lockheed Martin Corp. subsidiary Sikorsky's preaward protest of proposed Air Force solicitation language regarding acquisition of intellectual property), referring to Sikorsky Aircraft Corporation, B-416027, B-416027.2, May 22, 2018.

[8] 10 U.S.C. 2320(e).

[9] 10 U.S.C. 2320(e)(2).

later, when the OEM occupies a monopolist position as a sole source.[10] Just as a developer using options contracts has flexibility to acquire land for a reasonable price after obtaining a permit, the Air Force would be able to acquire technical data for a reasonable price after the need for such data arises.

Of course, this flexibility is not free because the Air Force must pay an options premium for the right to purchase data at a fixed price in the future. The premium amount may be more reasonable due to the presence of competition, but the Air Force must still determine whether the premium price is worth future flexibility. In doing so, a program office must weigh the likelihood that future activities requiring technical data will be needed, the estimated benefit of that activity, and the cost savings of paying the option price for requisite data, as opposed to the monopolist price the OEM would offer.

## Obstacles to Priced Options Contracts

Options contracts can mitigate many technical data challenges, and DoD has already begun to develop approaches to use options. One approach made possible by current acquisition regulations involves negotiating an advanced agreement under the EMD contract to limit the price that OEM could charge for technical data in the future.[11] However, this approach may not achieve service objectives because an advance agreement is not a free-standing contract. Rather, it enables parties to clarify the reasonableness and allowability of costs that may be incurred with respect to one or more principal contracts—in this case, the EMD contract.[12] Thus, it is unclear how an advance agreement could survive the expiration of the principal EMD contracts to meet technical data needs that arise many years into a weapon system's life cycle. Relatedly, even if an advance agreement sets the amount an OEM will charge for technical data, it does not independently impose any performance obligations. So the OEM may refuse to sell. Any performance obligation would reside with the principal EMD contract and would expire with that contract.

To create delivery contractual obligation that outlives the EMD contract, the Air Force would have to enter into a separate agreement with a longer duration: in other words, an options

---

[10] In practice, the option price is not likely to be a set figure. Rather, the parties may agree to a formula to determine the price of technical data. That formula can account for unpredictable variables, such as cost inflation.

[11] See Federal Acquisition Regulation (FAR) 31.109, which states, "[C]ontracting officers and contractors should seek advance agreement on the treatment of special or unusual costs . . . [t]o avoid possible subsequent disallowance or dispute based on unreasonableness, unallocability or unallowability under the specific cost principles [in FAR Part 31]." Although FAR 31.109 does not specifically mention IP development costs, its list of illustrative examples includes such things as "independent research and development" costs and "royalties and other costs for use of patents," suggesting that an advance agreement for data rights would be consistent with FAR 31.109.

[12] FAR 31.109.

contract. While an advance agreement may help clarify how the option price is calculated—for example, predefining cost allowability—it is by no means a substitute. However, several obstacles prevent program offices from fully utilizing options. We discuss three major obstacles below: (1) securing options for sufficiently long periods, (2) preserving funding authority to exercise options funds, and (3) providing program offices with proper policy guidance.

## *Insufficient Option Period*

The period covered by commercial options contracts typically ranges from several months to a few years. Crude oil options on the Chicago Mercantile Exchange, for instance, have expiration dates measured in months. But because many technical data needs do not arise until well into a weapon system's life cycle, a technical data purchase option must last decades to be reasonably useful.

In principle, there is no reason why a purchase option cannot span multiple decades.[13] However, DoD has placed regulatory barriers that make it nearly impossible for the Air Force to enter into such agreements. DFARS 217.204 specifies that any contract option may have a term of up to five years. While this limit can be extended, the period cannot last more than ten years without a finding by the agency head that "exceptional circumstances" warrant a longer term.[14] And DoD must report any options over ten years to Congress.[15] These requirements effectively prevent the Air Force from entering into options contracts that last more than ten years. Given that major weapon systems have multidecade life cycles, any conceivable options contract will have a duration that falls short of usefulness. As such, any program office following Congress's mandate to "address the merits of using a priced contract option" is likely to conclude that such options are meritless.

A congressionally mandated panel (Section 813 Panel) recently recognized that regulatory limits on contract options' lengths undermine the potential benefits of technical data options.[16]

---

[13] The five-year limit on multiyear contracts does not apply because options are not multiyear contracts. As Federal Acquisition Regulation 17.103 (2002) explains, "the key distinguishing difference between multi-year contracts and multiple year contracts is that multi-year contracts . . . buy more than 1 year's requirement (of a product or service) without . . . having to exercise an option for each program year after the first." FAR 17.103. See also *Freightliner Corporation*, Armed Services Board of Contract Appeals (ASBCA), 94-1 BCA 26,538, 1993 WL 502202, ASBCA No. 42,982 (November 26, 1993) (multiyear contract limitation did not apply to option quantities). Nor does the requirement under FAR 17.204(e) that "the total of the basic and option quantities shall not exceed the requirement for 5 years" apply, because the quantity covered by the option would cover only one year of requirements. See *Delco Electronics Corporation*, Comptroller General of the United States, B-244559, October 29, 1991, 91-2 CPD ¶ 391 (option extending contract to seven and a half years was permissible because option quantity did not exceed five years of requirements).

[14] DFARS 217.204.

[15] We are not aware of any instance in the past decade when any DoD agency found an "exception circumstance" to justify an over-ten-year option and reported the contract to Congress.

[16] See Section 813 Panel Report, Tension Point Paper 25, "Time Limits on Priced Contract Options," p. 2.

That panel recommended that DoD amend DFARS to allow priced contract options for technical data and data rights to last up to 20 years.[17] While the proposed 20-year limit on contract options is an improvement over the current situation, it may still fall short of the necessary maximal duration. The life cycles of major weapon systems are extremely long. For instance, the Air Force continues to fly the KC-135 Stratotanker, which was first produced over six decades ago, in 1955. And the F-35 Lighting II is expected to remain in service until 2070. The need for technical data and rights may arise at any point during long life cycles.[18] Thus, technical data options that last beyond 20 years may be appropriate in many cases. We agree with the 813 Panel that the current 10-year limit on options should be lengthened in the technical data context. But we do not believe a 20-year limit is sufficient, based on the longer lifetimes of many Air Force major weapon systems.[19]

*Preserving Obligation Authority*

If the Air Force exercises a purchase option for technical data or associated rights, it must pay the option price calculated using the agreed on formula. This requires the Air Force to have the appropriate obligation authority. Maintaining this authority year after year, however, may prove challenging. Each year, the Air Force is required to calculate and score the option price of its active options against its budget. And if the Air Force does not exercise the option, it may use the funds for alternative purposes. The temptation to divert options funds may be substantial. This is especially so because the need to renew obligation authority each year imposes administrative burdens that the Air Force may seek to avoid.

One potential way to preserve funding authority is to request such authority on an ad hoc basis. That is, once the Air Force identifies an activity that may require the exercise of technical data purchase options, it would make an appropriate budget request. This approach would require sufficient lead time in identifying a technical data need so that the funding authority may be granted in time to purchase the necessary technical data. Alternatively, the Air Force (or DoD) may establish and maintain a centralized fund that can be used to fund for the exercise of options on an ad hoc basis.

---

[17] Under the panel's recommendation, any option lasting less than 10 years would require approval "at least one level above the contracting officer," and an option of between 10 and 20 years requires "approval from the Head of Contracting Authorities." Section 813 Panel Report, Tension Point Paper 25, "Time Limits on Priced Contract Options," p. 2.

[18] For instance, the Air Force transitioned depot-level maintenance of the KC-135 from Boeing to organic in 2011—over a half century into the aircraft's life cycle.

[19] The Section 813 Panel also recommended amending 10 U.S.C. 2320 to establish a rule requiring technical data purchase options "not to exceed 20 years." We do not believe it is necessary or appropriate for Congress to set a specific maximum period.

DoD's current policy prohibits program offices from purchasing technical data or associated rights without first determining that such purchase is "necessary to satisfy agency needs."[20] This policy undermines the effective use of contract options. Contract options are inappropriate where the asset at issue is deemed necessary—the buyer would simply purchase the asset rather than an option. Contract options are useful only when the value of that asset falls short of necessity—for example, if the asset were useful only under certain conditions that may or may not occur. But the plain language of DFARS—and attendant policy documents—suggests that program offices *should not* acquire technical data or data rights under those circumstances. Changing this message is a necessary (but insufficient) step to the effective use of options contracts.

DoD need not amend DFARS—though it could do so if it were so inclined.[21] Rather, it would be enough for the Air Force to clarify that the term *agency needs* should be interpreted broadly to encompass long-term sustainment activities that may require access to technical data and rights. Such clarification would be consistent with Congress's instruction that program offices must consider long-term sustainment needs in making source selection decisions.[22] However, as discussed in Chapter 3, program offices are naturally inclined to prioritize near-term objectives ahead of long-term ones. The Air Force must therefore provide appropriate resources and incentives to enable and encourage program offices to develop acquisition strategies that properly consider the long-term value of technical data.[23]

## Determining the Price for Technical Data or Data Rights Delivered When the Air Force Exercises an Option

When the Air Force exercises an option of the type described above, what price should it pay for the deliverable? As the Section 813 report notes, extensive capability exists outside the Air Force to answer this question. And the academic and trade literatures on the value of IP are voluminous.[24] We can use these literatures to identify key elements of setting the price for

---

[20] DFARS 227.7103-1(a)

[21] DoD may state, for instance, that it remains inappropriate to purchase technical data that falls short of being "necessary to satisfy agency needs" but clarify that program offices may enter into options contracts for technical data that may satisfy potential agency needs.

[22] 10 U.S.C. 2320(e).

[23] See supra at Chapter 4.

[24] Section 813 Panel Report, Tension Point Paper 5, "Intellectual Property (IP) Valuation." For more information on the broader literatures, see Weston Anson and Donna Suchy, eds., *Intellectual Property Valuation: A Primer for Identifying and Determining Value*, Chicago, Ill.: American Bar Association Section of Intellectual Property Law, 2005; Harald Wirtz, "Valuation of Intellectual Property: A Review of Approaches and Methods," *International Journal of Business and Management*, Vol. 7, No. 9, May 2012, pp. 40–48; Len Smith, *Valuation of Intellectual Property*, ISYM 540, Current Topics in ISM, July 2, 2009; European IPR Helpdesk, "Fact Sheet: Intellectual Property Valuation," June 2015.

data-related deliverables. Although these literatures offer many different perspectives on how to define and measure this value, a few general principles appear repeatedly:

- An IP asset can be hard to define. To assess its value, it must be clearly enough defined to identify an owner with well-defined rights to it.
- The value of an IP asset at any point in time is the value at that date of the economic benefits that it will yield for its owner over the course of its remaining life.
- Three broad approaches are available to assess this value. Many variations exist to address valuation that is specific to circumstances, but all of them implement one of these approaches or a hybrid of them.
  - The *income* approach seeks the present value of the economic benefits expected to accrue to its owner over its remaining life.
  - The *market* approach seeks the value that a free, competitive market for the asset would establish if a free market for the asset existed. If such a market existed, no assessment of the value of the asset would be required. So the market approach seeks to simulate what value a free market would yield for the asset in question.
  - The *cost* approach seeks the cost of re-creating the asset or creating a substitute asset that would yield the same flow of benefits over its lifetime. The cost approach can be used only to assess the minimum value of the asset, since an investor would re-create an asset only if the asset was expected to yield higher benefits than the cost of creating it.
- All of these approaches yield estimates of value. These estimates can include a suitable confidence interval around any point estimate. No matter how sophisticated the details of any approach in a particular setting, important elements of the estimation are inherently subjective.
- The preferred approach depends on details about the asset being assessed. Assessors are often encouraged to apply more than one approach so that one estimate can be used to cross-check another.

A publicly accessible, consensus-based guide exists for valuing one kind of IP—a brand. This is International Organization for Standardization (ISO) Standard 10668:2010.[25] The guide makes it clear that similar guidance could be applied to valuing other types of IP—for example, a patent. But it formally endorses a consensus approach only for valuing brands. A commercial brand is clearly quite different in character from the IP that the Air Force associates with technical data or data rights.

---

[25] International Organization for Standardization, *Brand Valuation—Requirements for Monetary Brand Valuation, ISO 10668: 2010*, Geneva, Switzerland: International Organization for Standardization, 2010. This standard was last reviewed and confirmed in 2017. Therefore, this version remains current. Because this is a proprietary document, we cannot quote details about its content here.

The Section 813 Panel Report describes a second, ongoing effort to develop consensus guidance by the Licensing Executives Society USA and Canada. The report states that this guide is expected to become available in 2019. Based on the current literature and its application in the ISO standard that we discuss here, we expect this new guidance to be qualitatively similar to that in ISO 10668 but framed to address the value of a broader range of IP asset types. We expect the basic principles applied to this broader range to be close to those discussed here.

That said, this guide is consistent with all the statements about valuing IP above. It emphasizes that the value of an asset depends heavily on details of the setting around it. As a result, the standard requires assessors to describe (among many other things)

- the *purpose* of the value estimate
- the intended *audience* for this estimate
- the *concept* underlying the approach to estimation
- the *compatibility* of that concept and approach with the purpose of the estimate.

The guide names a wide range of potential purposes, including two that might be relevant in our setting—the value relevant to a legal transaction or a license. In our setting, the *purpose* of the valuation is to set a price for access to technical data or, potentially, a change in the data rights accorded to the government under DFARS.

The intended *audience* includes the contractor and the various Air Force offices that will examine this valuation during source selection and when the Air Force requests delivery of technical data or new data rights. This valuation is relevant to the creation of a new asset, not a new assessment of a preexisting asset. The creation of this asset will presumably generate benefits for the Air Force and costs for the contractor. The contractor's costs will include the cost of creating the asset and the effect of Air Force ownership of the asset on future contractor cash flows.

The *concept* relevant to this setting is the instantiation of the contract type chosen to govern any transfer of technical data or data rights from the contractor to the Air Force. We envision a contract formed as part of the EMD program—that is, concurrent with the EMD contract—and kept active over the life of the program using mechanisms like those described above. In this setting, the contractor offers terms for setting a price during the EMD source selection that the Air Force then applies when it exercises its option to take delivery of technical data or new data rights.

To buy technical data, it is natural to specify the data to be delivered in a CDRL associated with a CLIN that prices these data with a cost-plus-fixed-fee contract type. The cost in question is the allowable cost associated with (1) creating data that the contractor does not have to create to execute the EMD or (2) transforming any technical data described in the CDRL into the format and medium that the CDRL dictates. This concept is *compatible* with the cost approach described above and, in particular, with the variant of it that re-creates an IP asset. In our case, the asset is being created for the first time. DFARS and DoD cost standards provide details on what costs are allowable and how they should be documented. In this setting, allowable costs do not include costs to the contractor beyond those directly associated with creating the asset.

In our setting the EMD-concurrent contract should include CDRLs that detail any technical data that the Air Force wants to maintain an option to buy in the future. The detail should be fine enough to allow the Air Force to verify that it has received what it requested on delivery.

Presumably, the details of the contract type associated with the CLIN for each set of technical data would be the same.

Note that the allowable costs associated with a new asset cannot be assessed until the contractor creates the asset. That means that the EMD-concurrent contract sets the terms of the agreement for delivery of technical data when delivery occurs. But because the price will be set based on costs realized in the future, no escalation clauses are required in the EMD-concurrent contract. That said, to support an assessment of cost realism during the EMD source selection, the Air Force will need to state clearly how it will assess the costs that an offeror associates with future delivery of technical data. It may be worthwhile to bring such language to the EMD-concurrent contract itself to support Air Force assessment of allowable costs when delivery occurs in the future.

Things are more complicated if the Air Force seeks to change the assignment of data rights in the future. There are no traditionally allowable costs associated with such a change. If the Air Force wants to preserve an option to change data rights in the future, it presumably has two broad options.

One is to set a firm fixed price in the EMD-concurrent contract CLIN that defines this option for each change that might occur. That price might be indexed to reflect changes in the likely effects of such a change on the contractor over time. The range of changes that might occur in the future could be enormous, particularly if we imagine special negotiated license rights to be delivered in the future. The transactions cost of setting such prices during the EMD source selection is likely to be high. And the benefit could be small if the Air Force rarely exercises the options to change its data rights in in the future.

Alternatively, the EMD-concurrent contract could treat such changes as changes in contract scope that are subject to equitable adjustment in the contract price. In this case, the contractor will be able to demand compensation for the profits it expects to lose if the Air Force requests expanded data rights in the future. Now the income approach to asset valuation comes into play. But it does not emphasize the value of the IP asset to the Air Force. Rather, it focuses on the loss of value that the contractor experiences when the Air Force takes delivery of this new IP asset. The Air Force can use the income approach to conduct its own assessment of what it expects to gain from additional data rights. But our interest remains the appropriate adjustment in price, which reflects the contractor's well-being.

In this setting, can the Air Force benefit from requesting additional data rights if it must compensate the contractor for the losses the contractor experiences when the Air Force gains these rights? As we explain in Chapter 3, the answer depends on the relative discount rates that the Air Force and contractor use as they apply the income approach to asset valuation. The Air Force is more likely to benefit from such a change, the lower its effective discount rate is relative to the contractor's. Keep in mind that time is less important in this setting than in the discussion above. In Chapter 3, we discussed the Air Force making investments in technical data and data rights early in the life cycle of a program in expectation of receiving benefits decades in the future. This long delay between investment and benefit increased the importance of the difference between Air Force and contractor discount rates. In our current setting, we consider an Air Force decision to change data rights during the O&S phase to achieve benefits during the remainder of the O&S phase. There is no delay between the Air Force investment and the initial flow of benefits to the Air Force and costs to the contractor. This lack of delay requires a larger difference between the Air Force and contractor effective discount rates for the Air Force to benefit from such a change.

## Summary

Contract options offer a promising tool that enables the Air Force to turn away from its historical underinvestment in technical data. But to be truly effective, the Air Force must resolve several obstacles. First, the Air Force must find a way around DoD's prohibition against options that last more than ten years. Second, the Air Force must ensure that it preserves funding authority to exercise the option if the need arises. Third, the Air Force must clearly state how to measure the cost of exercising an option and translate this cost into a price. And finally, the Air Force should clarify to program offices that they may purchase technical data options to meet likely long-term sustainment needs and provide training on the use of such options.

# 8. Recommendations

Based on the work commissioned by the FY 2016 NDAA panels and studies, as well as the research developed during this and prior RAND projects, we developed the following recommendations for senior Air Force leaders seeking better acquisition and use of IP for major weapon systems.

## Structure Different Approaches to Intellectual Property Policy for Legacy and New Systems

If the Air Force could get effective access to IP for free, the most productive place to pursue IP would be in legacy systems. This is where the Air Force commits the majority of its O&S-related resources each year. Better access to IP could have the largest effects on O&S performance and costs in legacy systems. But it is not free to get access to IP. OEMs that support legacy systems are likely to suffer from lower profits if they transfer effective access to IP associated with these systems to the Air Force. The OEMs will agree to make such IP available only if the Air Force compensates them for their loss. For legacy systems, Air Force policy on IP should focus on what the Air Force can do in this constrained environment.

The Air Force is likely to get larger benefits from investing a dollar in IP in new systems than in legacy systems, particularly if it can make such investments before it becomes locked in with a sole-source OEM developer, producer, or sustainer of a major system. The Air Force can potentially benefit from such investments in requirements determination, acquisition strategy, market research, source selection, and contract formation associated with any new system. Effective investments in these stages can preserve the Air Force's flexibility later in a program's life cycle. Unfortunately, looking forward, such new systems will account for only a small share of the costs of O&S and sustaining the industrial supply chain for many years to come.

The Air Force should pursue changes in IP policy in both places. But the changes will differ between legacy and new systems. Within its budget each year, the Air Force should make the changes in legacy and new programs to yield the largest benefits for the Air Force as a whole. Its assessment of benefits, of course, will depend on the relative importance the Air Force places on short-term costs and long-term benefits.

## Understand the Basic Role of Time When Comparing the Cost and Benefits of Investing in Intellectual Property

The primary benefits of having effective access to IP come during the O&S phase of a program. The best way for the Air Force to get access to these benefits is to make a series of investments early in the life cycle of a program. But it is cost-effective to invest in this way only if the Air Force has the patience to wait for benefits that will come many years following any investment.

In any budget year, the Air Force must decide how to spend its available budget. The lower the Air Force's *hurdle rate* for investing in IP, the easier it is to justify allocating a portion of the current budget to such an investment. A hurdle rate is the minimum rate of return that an investment must promise to be eligible for funding when an organization allocates its investment budget. The corresponding *real* hurdle rate for an OEM is likely to be around 7 percent a year when corrected for inflation. If the Air Force hurdle rate is lower than this, two things help favor Air Force investments in IP.

First, when the OEM transfers IP to the Air Force, the gains the Air Force realizes are likely to exceed the costs to the OEM. This difference creates potential gains from trade that allow the Air Force to pay the OEM more than retention of the IP is worth to the OEM. If the Air Force hurdle rate exceeds that of the OEM, it will be more difficult to identify such gains from the trade and the potential they create for the Air Force to secure access to IP.

Second, in interactions with the OEM early in a program, the Air Force will be willing to invest relatively more resources and effort in these interactions than the OEM. As a result, the Air Force is more likely to realize favorable outcomes from these interactions. If the Air Force hurdle rate exceeds that of the OEM, the opposite will be true. The OEM will be more willing to invest in interactions than the Air Force and, as a result, will likely benefit more from these interactions than the Air Force.

In any case, the Air Force must invest persistently in activities relevant to IP to achieve any useful level of effective access to IP. Senior leaders must determine whether such persistent investment is worthwhile. If it is not, the recommendations that follow are irrelevant; they assume that such investment is worthwhile. If investment is worthwhile, senior leaders must act as persistent advocates for such investment, wherever it is appropriate to invest in the Air Force. We say more about where to invest below.

## Support More Proactive Pursuit of Effective Air Force Access to Intellectual Property in Major Programs

Many opportunities exist for the Air Force to invest resources and effort in improving the quality of the IP that it can access. Program offices typically do not pursue these in part because current Air Force policies discourage them from doing so. Examples we have discussed include challenging contractor assertions about data rights, challenging markings on drawings, challenging definitions of OMIT or FFF data, and ensuring delivery of technical data described in CDRLs in a timely manner. In such cases, program personnel

- are not aware that an opportunity exists, because they have not been trained about this opportunity

- are not aware that an opportunity exists, because they lack access to a relevant expert or are reluctant to engage an expert because that could send a message that things are not in control in the program
- are aware of an opportunity but lack the time or resources to exploit it or, equivalently, are encouraged to commit their limited time and resources to other issues
- are aware of an opportunity but are afraid that pursuing it will disrupt the program in ways that increase immediate program cost or delay its schedule.

Any of these things can occur when a more proactive stance could improve outcomes in the relevant program or, even it if complicated things in that program for some short period, could have positive consequences for the Air Force as a whole over the longer term.

Formal policies can drive some of these behaviors. Those can be mitigated by providing better access to training and expertise. They can be mitigated further by rewarding personnel for getting training and using outside expertise. The Air Force has a wide range of mechanisms for motivating people to behave differently. But they must be brought to bear in a coherent manner across many specific circumstances.

One very direct and visible policy is one that commits additional resources to a program office, especially if the resources are targeted to address IP-related issues. Such a policy requires a recognition that it is acceptable to increase the cost of a program today as an investment in better IP in the future. Such a policy can be sustained only if the senior leadership maintains visible and persistent support for this position.

Informal policies are often more difficult to change, in part because they are difficult to identify. An aversion to disrupting a program may become so second nature that no one recognizes the potential benefit of disruption—for example, provoking a confrontation with a contractor to get a better definition of some element of IP law in a courtroom. Program personnel will not take this on until they see others rewarded for taking such a position when it is appropriate.

Formal change management, discussed below, can help implement and sustain such changes in policy.

## Clarify the Scope of Operation, Maintenance, Installation, or Training and Form-Fit-Function Data

The technical data statute and DFARS grant the government unlimited rights to OMIT and FFF data, even when such data were developed exclusively using private funds.[1] But the rules also state that OMIT rights do not extend to "detailed manufacturing or process data."[2] The definitions of these terms do not provide sufficient clarity and have resulted in disputes. For instance, the government is interested in interpreting OMIT broadly to capture unlimited rights to

---

[1] 10 U.S.C. 2320(a)(2)(C); DFARS 227.7103-5(a)(5).

[2] 10 U.S.C. 2320(a)(2)(C); DFARS 227.7103-5(a)(5).

a large set of data, while industry favors a broad interpretation of "detailed manufacturing and process" to narrow the scope of OMIT.[3]

Several panel reports have recommended that DoD issue a guide or appendix defining the disputed terms with greater clarity.[4] However, such guidance is unlikely to resolve disputes because it would carry little to no legal weight. While an agency's reasonable interpretation of its own regulations is normally entitled to legal deference, that deference is not appropriate where the regulation merely paraphrases the authorizing statute.[5] This is because, without a more specific regulatory definition, the agency would be issuing an interpretation of the statute without undergoing the requisite rule-making process. Here, DFARS does not define OMIT at all, and its definition of "detailed manufacturing or process data" is merely a verbose rephrasing of the statutory term.[6] As such, an interpretative guide for these terms is unlikely to be legally binding.[7] DFARS does define FFF in greater detail than the statute and so may be able to offer authoritative guidance within the bounds of that definition.[8]

DoD may nonetheless issue its own interpretation of OMIT and "detailed manufacturing or process" data by amending DFARS to contain its preferred definitions. Congress directed DoD to "prescribe regulations to define the legitimate interest of the United States . . . in technical data."[9] Current regulations do not appear to adequately define the government's "legitimate interest" in OMIT data. An obvious remedy is to "prescribe regulations" to do so. Issuing new regulations is not without drawbacks. The process may be burdensome and time consuming, particularly if industry pushes back with political pressure. And even the most detailed regulatory definitions cannot possibly cover every potential circumstance, so ambiguities and disputes regarding the scope of OMIT likely will remain.

---

[3] See Section 813 Panel Report, Tension Point 16.

[4] See Section 813 Panel Report, Tension Point 16 ("recommends a guide with definitions"); Van Atta et al., 2017 ("develop an appendix to the DFARS that would specify in greater detail the meaning of such terms as "operation, maintenance, installation and training" data; "form, fit and function" data; and "detailed manufacturing and process" data).

[5] This deference, however, is subject to substantial criticism from scholars and jurists. See, for example, *Talk America v. Mich. Bell. Tel.*, 131 S. Ct. 2254, 2266 (2011) (Scalia, J., concurring) ("seems contrary to fundamental principles of separation of powers to permit the person who promulgates a law to interpret it as well"). *Gonzales v. Oregon*, 546 U.S. 243, 257 (2006).

[6] DFARS 227.7013-5(a), defines detailed manufacturing or process data as "technical data that describe the steps, sequences, and conditions of manufacturing, processing or assembly used by the manufacturer to produce an item or component or to perform a process."

[7] An interpretative guide may nonetheless be helpful if DoD finds that it lacks a coherent internal definition of OMIT.

[8] DFARS Part 227; DFARS 252.227-7013. FFF data as "technical data that describes the required overall physical, functional, and performance characteristics (along with the qualification requirements, if applicable) of an item, component, or process to the extent necessary to permit identification of physically and functionally interchangeable items."

[9] 10 U.S.C. 2320(a)(1).

DoD may also rely on dispute resolution mechanisms, including litigation, to resolve ambiguous terms such as OMIT. Indeed, Congress has prescribed a dispute resolution system within 10 U.S.C. 2321: DoD may reject a contractor's attempt to restrict what it believes to be the government's legitimate OMIT or FFF rights, and the contractor may challenge that decision before a judicial or administrative body. Although costly, litigation may nonetheless be worthwhile in the long run because courts apply rules on a case-by-case basis, and those rulings clarify the scope of OMIT or FFF for future parties. But litigation may be risky and requires DoD to accurately identify and reject meritless contractor assertions. The contractor is often the "last mover" in determining whether to proceed to litigation—it could back down before a DoD adverse determination and convey disputed OMIT data with unlimited rights if facts are unfavorable. Conversely, contractors are more willing to litigate under favorable facts, so they can obtain ruling that supports their legal interpretations—at the expense of DoD's positions.[10]

In sum, DoD may rely on interpretative guides to resolve ambiguities where possible. But that is likely not possible with respect to certain terms (including OMIT) that are not defined in DFARS. In such cases, DoD may amend DFARS to provide preferred definitions, but that may trigger political pushback. DoD may also attempt to advance its interpretation of OMIT through litigation, but that carries the risk of unfavorable rulings.

## Develop an Intellectual Property Cadre and Motivate Its Effective Use

The FY 2018 NDAA mandated creation of a cadre of specialists in IP but provided little specific guidance on what that cadre should look like, where it should reside (whether OSD level or service level), or how the Air Force should use it if the latter. We offer the following recommendations for the design and use of such cadres by the Air Force.

Align cadres to the way the Air Force normally acquires and sustains its major systems. This suggests that a cadre should reside in each system center to support the program offices in that center. High-level policy oversight of these cadres should lie in SAF/GC and SAF/AQ; day-to-day execution of policy should occur in the centers. Members of each cadre should not be matrixed to individual program offices for extended periods. Rather, they should be easily accessible by program offices during the episodic periods when the program offices can use a cadre's expertise to make a decision.

Include IP attorneys, O&S specialists, and personnel familiar with how Air Force program offices manage IP issues as part of each cadre. Gathering such specialists at system centers can create enough critical mass for members of each functional area to stay current on developments within its own function. But each center should strive to build cross-functional skills in its cadre

---

[10] To minimize this risk, DoD may attempt to accept the contractor's restrictive assertion in the middle of a case so as to moot the dispute and deprive a court of jurisdiction.

so that its experts can provide integrated support to program offices when a cadre supports their decisionmaking.

Serve as advocates for and core participants in development of IP policy at key milestones in major programs within a system center. Cadre participation should be mandatory in programs of sufficient dollar size or, at the secretary's discretion, importance. Specialists should play core roles in requirements determination, development of an acquisition strategy, market research, design of a source selection, contract formation, any major renegotiation of IP arrangements over the course of EMD, and comparable activities during an upgrade, modernization, or life extension during O&S.

Develop and provide group training for relevant personnel within a system center. Provide just-in-time training for activities entering predictable events over the course of a program life cycle—for example, before requirements determination, acquisitions strategy, market research, source selection, EMD execution, O&S execution, and supply chain sustainment. Following training, such personnel could support cadre personnel in major programs or execute appropriate responsibilities in lower-priority programs.

Work with the Air Force Institute of Technology and Defense Acquisition University to develop modules for individual training on IP issues for nonspecialists who expect to address IP issues in their programs.

Set standards for group and individual training. Maintain a record of individual participation in training, as an individual or in a group. Advocate for the resources required to sustain appropriate group and individual training.

Build and maintain access to an archive of best practices relevant to the kind of IP present in each system center. Each cadre should track best practice outside the Air Force and organize information on recent IP-related activities within each center that its program offices can benefit from. Liaison personnel should sustain horizontal links between system centers and with IP specialists outside the Air Force.

Maintain regular dialogue with IP specialists in defense contractors that play important roles in major Air Force programs in each system center. Participate in their trade associations and training activities. Where possible, seek common ground on issues and exploit that common ground quickly to realize mutual gains. Identify more contentious issues for more in-depth or longer-term consideration. Use these dialogues to improve contractor understanding of Air Force IP priorities and contractor options for addressing those priorities.

Work with IG to monitor compliance with Air Force policy on IP within each system center. Validate the template IG will use to monitor compliance in each regularly scheduled inspection. Provide appropriately trained personnel to support these inspections.

Notwithstanding the statutory authority to use DoD funds for the first three years of the cadre, the Air Force should eventually resource the cadre in each system center with core system center funds. Define and sustain a standard level of service that each program office in a system center can expect to receive without paying for it from the center cadre. Size each cadre to the activities assigned to the cadre (for example, those above), the level of service, and the expected demand on the cadre from ongoing and upcoming programs within each system center. Maintain additional capability and capacity to support programs willing to pay for services beyond the standard level of service from their own funds.

As discussed below, the Air Force should continue research on the IP elements of its acquisition workforce, to identify how best to recruit, retain, and manage this critical talent element. The Air Force should also monitor the IP cadre's work closely, to ensure effective resourcing and monitoring of cadre activities to learn when IP issues arise and assess the return on investment in IP data management.

## Create a Standard Mechanism to Preserve an Option to Acquire Intellectual Property at a Price Negotiated as Part of the Engineering and Manufacturing Development Source Selection

An option would allow the Air Force and OEM to agree in advance on the terms for future delivery of IP through the full life of a new program. The agreement would cover four issues:

1. what data the Air Force could request for delivery in the future, fully specified in terms of content, level of detail, format, and the medium in which the Air Force would use the data
2. what rights the Air Force would have in these data when they were delivered in the future
3. what premium price the Air Force would pay the OEM up front to create and preserve this option
4. what option price the Air Force would pay the OEM at the time of delivery.

Such an option would go to the heart of the challenge the Air Force faces in acquiring IP. The Air Force could negotiate the terms of delivery during a source selection in which competition determined the terms of offers to deliver data in the future on demand. The Air Force could then wait until it had better information on what data it needed to request delivery. Properly designed, such an option would limit what data the Air Force feels compelled to acquire in the EMD source selection. And it would limit the price the Air Force paid for data that it determined it would need.

The contract for an option would be drawn up concurrently with the EMD contract but would be separate from it. This would allow the options contract to remain in force beyond the final date on which the OEM could deliver products of the EMD effort. To last more than 10 years without a special report to Congress, this approach to contracting would require a change in DFARS 217.204. The Section 813 Panel recommended an extension to 20 years. To ensure that the Air Force and OEM could agree on terms in advance that would remain in place for the life of a program, the allowable term on the contract would need to extend well beyond 20 years.

When the Air Force has an option that it can exercise in a particular year, it must program the funds it would need to exercise the option. If these funds are not used to exercise the option, they can be applied elsewhere. But programming the funds and then redirecting them imposes transaction costs that could be reduced by allowing more flexibility in programming. Many mechanisms are possible. A simple one would be a general fund programmed each year that the Air Force could use to exercise any option in that year. When the general fund was exhausted in any year, the Air Force would no longer be free to exercise additional options in that year.

Setting a price for technical data will be easiest if the price is set on delivery. To avoid exploitation by a sole source at that time, the Air Force and OEM could agree in advance that the price at delivery would cover allowable costs associated with the creation of the data requested. The challenge here is how to define what costs would be allowed and how they would be measured and audited. One benefit of creating a single standard options contract is the need to define such allowable costs only once.

The Air Force should not make use of the standard contract mandatory. Rather, the standard contract should be available for an Air Force program office to use if it so chooses. The office could decide that no mechanism was needed to acquire additional IP in the future. It could decide to negotiate its own SNLR.

## Coordinate Intellectual Property Policies and Execution Across All Phases of Acquiring a Program

When setting priorities among recommendations like those above, the Air Force should keep in mind that these recommendations are designed to work together as a package. Emphases will be different in programs for legacy and new systems. Given the role of time in IP policy, the Air Force leadership must decide whether it is willing to make a sustained commitment to make the investments necessary to acquire IP the Air Force might use in the future. If the leadership decides it is worthwhile for the Air Force to pursue IP over the long term, it must then commit to doing this in each phase of a program. For example, it must make the following kinds of complementary investments:

- Determine the requirement for IP. If flexibility to allow adjustment in a sustainment plan well into the life of a program is important, the Air Force should ensure that that is stated as a requirement.
- Work with potential offerors to determine what IP the Air Force would need to maintain any life-cycle plans for O&S, modernization, life extension, supply chain sustainment, and so on.
- Craft the CDRLs relevant to IP and the CLINs that will be used to price them, and penalize any failure to deliver.
- Determine how to assess the risks of shortfalls in IP included in any offer.
- Validate contractor assertions of data rights, appropriate markings, and delivery of data.
- Protect IP as trade-offs are made during EMD in response to the outcomes of ongoing development.
- Record agreements on IP in a form that ensures that a program office remains well informed on agreements made in the past.

In any program, successful implementation of each of these actions increases the productivity of the other actions. These actions are complements more than substitutes for one another.[11] If the Air Force does not have enough resources to do all of these things, it should choose programs in which to pursue IP and focus on all of these actions within a program rather than focusing on one of these actions across programs.

All of these actions will be more productive if the Air Force creates a cadre that can act as a general advocate for IP policy in the Air Force–wide budgeting process, in the training of the personnel that will execute all of these actions, and in the coordination of these actions within any particular program.

## Use Formal Change Management to Coordinate the Diverse Activities Necessary to Improve Intellectual Property Policy

Formal change management is a set of practices that large, complex, technologically sophisticated organizations have used to implement changes that affect many parts of the organizations and require alignment of all these parts to a common approach to change. Formal change management is well suited to the task of improving IP policy in an organization like the Air Force. Many diverse functional communities are involved. Change must occur over a period and then persist over a still longer period, even as key leaders in the organization turn over every few years. Many elements of formal change management are analogous to those the Air Force already uses routinely to manage its acquisition of major systems.

Formal change management proceeds incrementally. It translates high-level, enterprisewide goals into specific targets in pilot programs that can be conducted to familiarize a large organization with a new way of doing business. Pilots provide laboratories in which the organization can test new ideas and adapt them to the organization's needs. As successful implementation of change occurs in pilots, information on this success is used to justify committing additional resources and effort to expanding the change effort until it encompasses the whole organization.

Formal change management requires the coordination of all functions in which personnel must change their behavior for organizational change to succeed. It prepares for change by creating new capabilities and designing actionable plans for closely instrumented pilots. It executes changes within these pilots. And then it institutionalizes these changes so that processes in place before can be removed.

Throughout these stages of change, it resources and trains those in the change effort. It uses all the mechanisms consistent with the organization's culture to convince people who must

---

[11] The principle underlying this point is explained in Paul Milgrom and John Roberts, "The Economics of Modern Manufacturing: Technology, Strategy, and Organization," *American Economic Review*, Vol. 80, No. 3, June 1990, pp. 511–528.

change their behavior for organizational change to succeed that they will be better off if they change their behavior than if they do not. It monitors their performance. And it continually communicates the leadership's goals for change, the challenges arising during change and mitigations that have successfully addressed them, and evidence of how successful change has improved the organization's performance and the well-being of those who have changed their behavior.

The Air Force has successfully used this kind of formal change management in the past. And resources are available to help it do so in the future.

## Continue Research and Policy Development Relating to Intellectual Property Strategy, Acquisition, Use, and Management by the Air Force

This project identified future research vectors for the Air Force to consider in the field of IP strategy, acquisition, use, and management, as well as such supporting fields as acquisition workforce development and management. Future research questions raised by this project include the following:

- How can the Air Force effectively recruit, retain, and manage its acquisition workforce, including high-demand niche specialties such as lawyers or contracting personnel with expertise in IP acquisition and management?
- What are the knowledge, skills, and abilities the Air Force needs in its IP cadre, and how well do Air Force education and training institutions impart these knowledge, skills, and abilities to Air Force personnel? Are there alternate strategies (such as outsourcing or use of civilian education) that could help the Air Force develop its IP expertise?
- How can the Air Force measure the rate of return on IP expertise or measure the rate of return for IP acquisition in general? Is it possible to discern the most critical or valuable types of IP expertise or rights for future investment?
- What can the Air Force learn about its future IP needs and issues from IP management on major weapon systems currently in development or production? Can the Air Force clarify IP rights at this stage such that it will have more flexibility or certainty with respect to future use of IP for operation, sustainment, and modification of these systems?
- What alternate bases exist for the allocation of IP rights, beyond the current framework of FAR and DFARS that allocates rights based on funding sources? What effects might alternate allocation schemes have on pricing and contractor behavior?
- What structures can be established for the acquisition and management of IP developed at government expense (either through contractors or by government labs and personnel)? How should the government manage this IP portfolio and rights therein? What legal and policy innovation or change is necessary to support this portfolio?

# Appendix A. Findings of the Studies Mandated by the Fiscal Year 2016 National Defense Authorization Act Sections 809, 813, and 875

In Sections 809, 813, and 875 of the FY 2016 NDAA, Congress commissioned three separate reviews of acquisition laws and policies that related, in some way, to IP. The panels and studies commissioned by the FY 2016 NDAA each took a different approach to their work. Sections 809 and 813 directed the creation of panels that included public- and private-sector representatives to study acquisition issues. The Section 809 and Section 813 Panels organized themselves and focused on the questions posed by their legislative mandates, producing reports that differed substantially in both form and substance. By contrast, based on the language of the Section 875 mandate, OSD commissioned IDA to respond directly to the questions posed by Congress through the use of a study. This appendix summarizes the relevant work by each panel or study, which we reviewed in the course of our analysis for this project.

## Section 809 Panel and Reports

Congress commissioned the Section 809 Panel in the FY 2016 NDAA to study a broad array of issues relating to the streamlining of defense acquisition and procurement issues. This charter was subsequently amended by Congress twice, by the FY 2017 and FY 2018 NDAAs, to provide for administrative support and other matters. The Section 809 Panel was required to include "at least nine individuals who are recognized experts in acquisition and procurement policy," with congressional direction to DoD to ensure that "the members of the panel reflect diverse experiences in the public and private sectors."[1] Congress gave this panel the mandate to

(1) review the acquisition regulations applicable to the Department of Defense with a view toward streamlining and improving the efficiency and effectiveness of the defense acquisition process and maintaining defense technology advantage; and

(2) make any recommendations for the amendment or repeal of such regulations that the panel considers necessary, as a result of such review, to—

    (A) establish and administer appropriate buyer and seller relationships in the procurement system;

---

[1] Section 809 of the National Defense Authorization Act for Fiscal Year 2016 (Public Law 114-92), as amended by section 863(d) of the National Defense Authorization Act for Fiscal Year 2017 (Public Law 114-328).

(B) improve the functioning of the acquisition system;

(C) ensure the continuing financial and ethical integrity of defense procurement programs;

(D) protect the best interests of the Department of Defense; and

(E) eliminate any regulations that are unnecessary for the purposes described in subparagraphs (A) through (D).[2]

The first volume of the Section 809 Panel Report addressed IP issues, focusing on how technical data rights requirements affect the government's ability to acquire commercially available and commercial off-the-shelf items. The panel report stated that the current definition(s) used to determine the commerciality of an item may possibly exclude commercially available items that are not widely available, particularly those items that are cutting edge and therefore in limited production. Furthermore, the panel indicated that the government should not apply government-purpose rights to those items that have minor, noncommercially available modifications made at the government's insistence.

Several parts of the first volume described the tension between commercial practices and government practices or requirements. It stated that

> [the government's] unique definition [of rights in data] sets up a number of conflicts and adds confusion over the critical issue of proprietary data restrictions for products procured in the commercial market. . . . It is also unclear if an item with minor modifications incorporated to meet unique federal government requirements is likely to be sold in substantial quantities to the general public.[3]

The Section 809 Panel's first volume also argued that current technical data rights requirements for commercial items deter commercial businesses from doing business with the federal government.[4] In its most direct section on IP issues, the Section 809 Panel described a fundamental tension between commercial practice and government regulations, exemplified by distinctions between the FAR and DFARS language on data rights:

> Although the FAR itself emphasizes relying on customary commercial practice, the DFARS treats IP rights acquisition as an act of the sovereign, which can grant or deny as it pleases by claiming government purpose. In the commercial world, assignment of IP rights is subject to negotiation, and transfer of rights typically involves fair compensation in the form of a purchase or payment of license fees. The FAR is more closely aligned to commercial practice.[5]

---

[2] Section 809 of the National Defense Authorization Act for Fiscal Year 2016 (Public Law 114-92), as amended by section 863(d) of the National Defense Authorization Act for Fiscal Year 2017 (Public Law 114-328).

[3] Vol. 1, Sec. 809 Report, p. 25.

[4] Vol. 1, Sec. 809 Report, pp. 46–49.

[5] Vol. 1, Sec. 809 Report, pp. 46–49.

Volume 2 of the Section 809 Panel Report focused on the acquisition workforce, commercial source selection, the Cost Accounting Standards Board, and services contracting, as well as an "enterprisewide portfolio management structure."[6] The report mentioned IP in a few places, including one mention of "intellectual property rights issues that inhibit sustainment."[7] The same section describes the challenges that arise in connection with forecasting and acquiring the right quantity and type of IP rights to support future operation and sustainment of a major weapon system:

> Acquisition of data rights as part of weapon systems development has changed in recent years. If a weapon system is to be sustained through a combination of commercial and organic support, access to intellectual property (IP) rights that allow component repair—and in some cases, competition to provide those capabilities—is crucial. Appropriate planning, funding, and contracting for government acquisition of necessary IP is best accomplished up front, not as an afterthought. Requesting a complete data package might not be cost effective either. Instead, the government should consider obtaining rights to those specific portions and for the specific purposes (e.g., organic depot maintenance/acceptance testing versus detailed design/material data required to enable a future reprocurement of the component) of the system it foresees acquiring in the future.[8]

When Volume 3 addressed IP issues, it echoed a common industry concern that enforcement by DoD contracting officers of the government's technical data rights could hamper innovation and competition within the sphere of defense acquisition. The volume also emphasized the current lack of focus on data rights, indicating that these issues are often best addressed at the beginning of the acquisition process when competitive pressure is greatest and the government arguably has the greatest bargaining leverage. The Section 809 Panel Report also recognized that both sides have little knowledge about the future and what rights may be needed. However, the panel expressed confidence that these issues could be better addressed if more time were devoted to these issues (vice schedule, performance, or other considerations). Accordingly, the panel recommends that government officials focus on the quality of data rights to address relevant issues rather than the quantity of data rights acquired.

In support of these points, the Section 809 Panel Report paraphrased stakeholder interviews as reporting that "the intellectual property (IP) of private-sector companies will not be protected

---

[6] Cover letter, Vol. 2, Sec. 809 Report, June 28, 2018.

[7] Vol. 2, Sec. 809 Report, p. 48.

[8] Vol. 2, Sec. 809 Report, pp. 49–50.

by DoD under existing policy."[9] This sentiment was particularly acute with respect to DoD's expressed desire to acquire software source code:

> [A] number of software companies, and the private investors many of the companies rely on, fear that if DoD partnered with the company to develop an innovative solution, it could lead to DoD taking the idea and turning it into an RFP to find someone who might be able to produce the solution at a cheaper price.[10]

In addition to concerns about how the government might do harm through its acquisition of contractor IP, and the reverse engineering thereof, the third volume of the Section 809 Panel Report addressed the problems created during sustainment by inadequate IP acquisition. "IP and data rights are not appropriately addressed" by government program offices or contracting personnel, resulting in problems downstream for major weapon systems.[11] The Section 809 Panel further described the problem as beginning at the start of procurement but creating problems throughout the life cycle of major weapon systems:

> To maintain competition throughout the lifecycle, data rights and IP—as applicable to both hardware and software—must be addressed up front. Obtaining IP and data rights has become a complex issue for most major programs, resulting in dissatisfaction within both the organic and commercial depot organizations. Data rights and IP should be made available when needed, where needed, and for the specific purpose needed while also protecting the IP and data rights of industry partners.[12]

However, beyond better IP strategy at the front end of procurement, the report does not offer additional analysis or recommendations to help government or industry resolve the problems relating to IP that may arise over the life of a major weapon system.

Volume 3.2 of the Section 809 Panel Report discussed options and SNLRs. This volume emphasized the necessity of streamlining the documents needed in the acquisition process, including the use of commercial practices to the maximum extent practicable. Most notably, while this volume emphasized the needs of program offices to keep long-term sustainment costs low for projects, it does not address the acquisition of technical data rights as a potential option to lower long-term sustainment costs through the internalization of maintenance and upgrade capacity. In its Table 7-4, the panel outlined the current Acquisition Plan Contents and asserts the content of this plan are excessive. The current contents include the IP strategy for the acquisition. The report states that the acquisition plan should be reduced to the minimum

---

[9] Vol. 3.1, Sec. 809 Report, p. 26.

[10] Vol. 3.1, Sec. 809 Report, p. 26.

[11] Vol. 3.1, Sec. 809 Report, p. 111.

[12] Vol. 3.1, Sec. 809 Report, p. 112.

number of items that are "required by statute or truly critical" to the acquisition program but is unclear whether IP would be included within these documents.[13] The panel also pointed to Other Transaction Authority agreements as a potential panacea for solving complex problems associated with acquisition and procurement but is unclear how such agreements might be specifically used to better acquire IP.[14]

## Section 813 Panel and Reports

In Section 813 of the FY 2016 NDAA, Congress directed DoD to establish a "government-industry advisory panel" on rights in technical data.[15] The Section 813 Panel's mandate was to "review sections 2320 and 2321 of title 10, United States Code, regarding rights in technical data and the validation of proprietary data restrictions and the regulations implementing such sections, for the purpose of ensuring that such statutory and regulatory requirements are best structured to serve the interests of the taxpayers and the national defense." The panel's legislative mandate outlined the following scope for this effort:

> (3) SCOPE OF REVIEW—In conducting the review required by paragraph (1), the advisory panel shall give appropriate consideration to the following factors:
> (A) Ensuring that the Department of Defense does not pay more than once for the same work.
> (B) Ensuring that Department of Defense contractors are appropriately rewarded for their innovation and invention.
> (C) Providing for cost-effective reprocurement, sustainment, modification, and upgrades to Department of Defense systems.
> (D) Encouraging the private sector to invest in new products, technologies, and processes relevant to the missions of the Department of Defense.
> (E) Ensuring that the Department of Defense has appropriate access to innovative products, technologies, and processes developed by the private sector for commercial use.[16]

The FY 2016 NDAA directed DoD to establish the Section 813 Panel (as with the Section 809 Panel) with a mix of representation from government and industry. The panel's chair was selected by the then Under Secretary of Defense for Acquisition, Technology, and Logistics, and the

---

[13] Vol. 3.2, Sec. 809 Report, pp. 405–406.

[14] Vol. 3.2, Sec. 809 Report, pp. 440–441.

[15] Sec. 813, FY 2016 NDAA.

[16] Sec. 813, FY 2016 NDAA.

panel was filled with government members who "are knowledgeable about technical data issues and appropriately represent the three military departments, as well as the legal, acquisition, logistics, and research and development communities in the Department of Defense" and private-sector members who "include independent experts and individuals appropriately representative of the diversity of interested parties, including large and small businesses, traditional and non-traditional government contractors, prime contractors and subcontractors, suppliers of hardware and software, and institutions of higher education."

In its report, published in late 2018, the Section 813 Panel gave a short description of the IP problems arising during the acquisition of major weapon systems and then a series of *tension point papers* regarding discrete issues within this sphere. Some of these tension papers offer solutions or recommendations to the tensions so identified; others merely describe the problem and the respective positions of government and industry. In many cases, the tension points identify issues and potential solutions that are favorable to either government or industry but do not recommend a solution or resolution to the tension(s) identified. The Section 813 Panel's papers are listed in Table A.1.

### Table A.1. Section 813 Panel Tension Point Papers

| Tension Point Paper Number | Topic |
| --- | --- |
| Tension Point 1 | Different business models in government and industry result in different objectives |
| Tension Point 2 | Access for limited purposes (cyber review, airworthiness, approvals) versus delivery as a CDRL |
| Tension Point 3 | CDRL requirements for budget activities 6.1 and 6.2 research programs versus CDRL requirements for production/sustainment |
| Tension Point 4 | Data rights as an evaluation factor |
| Tension Point 5 | IP valuation |
| Tension Point 6 | Contract requirements in Section H should not be treated like standard clauses |
| Tension Point 7 | Treatment of independent R&D versus self-funded R&D for IP rights determinations; independent R&D risk correct for limited/restricted rights? |
| Tension Point 8 | Is source of funding the best way to determine rights to technical data? |
| Tension Point 9 | Commercial versus noncommercial items |
| Tension Point 10 | Commercial software license versus government-unique requirements |
| Tension Point 11 | Authorized release and use of limited rights technical data |
| Tension Point 12 | Are existing rights sufficient for maintenance and sustainment? |
| Tension Point 13 | Software versus technical data |
| Tension Point 14 | Development versus adaptation (includes majority and minority paper) |
| Tension Point 15 | OMIT data versus detailed manufacturing or process data, including such data pertaining to a major system component |
| Tension Point 16 | Rigid IP requirements versus flexible arrangements |
| Tension Point 17 | Poor data item description alignment with statutory/regulatory categories [i.e., FFF and OMIT] |

| Tension Point Paper Number | Topic |
|---|---|
| Tension Point 18 | The validation process is cumbersome and confusing (includes a narrative report and line-in/line-out report) |
| Tension Point 19 | Mandatory flow-down requirements (i.e., requirements for commercial subs and suppliers) (includes majority and minority paper) |
| Tension Point 20 | How to keep CDRL deliverable(s) up to date |
| Tension Point 21 | Small business innovation research (i.e., flow-down requirements to suppliers, inability to share with prime contractors, evaluation) |
| Tension Point 22 | Lack of trained personnel |
| Tension Point 23 | The data assertion list is a burden on both contractor and the government |
| Tension Point 24 | Deferred ordering period: six years (rather than perpetual) and deferred ordering for only data generated under the contract (includes majority and minority paper) |
| Tension Point 25 | Time limits on priced contract options |
| Tension Point 26 | Deferred ordering: all interface or major systems interface data may be ordered regardless of U.S. government development |
| Tension Point 27 | Failure to define and order CDRLs (reliance on deferred ordering and data accession lists to obtain data) |
| Tension Point 28 | Escrow as a form of deferred delivery |
| Tension Point 29 | Government purpose rights in major system interface (developed exclusively at private expense or with mixed funding) |
| Tension Point 30 | Government purpose rights in interfaces developed with mixed funding |

NOTE: This list is taken from Appendix D of the Section 813 Panel's final report, with minor edits to conform the listed title for each tension point paper to RAND style and usage.

## Section 875 Study

Section 875 of the FY 2016 NDAA called for a review of DoD practices regarding "access to and use of intellectual property rights of private sector firms" and "use of intellectual property rights to facilitate competition in sustainment of weapon systems."[17] OSD responded to this requirement by commissioning a report by IDA (the Section 875 Report).[18] The Section 875 Report found that gaps in access to technical data and associated rights have diminished competition for weapon systems sustainment and undermined DoD's ability to meet certain legal sustainment requirements.[19] The Section 875 Report offered six recommendations, reproduced in summary form below.

---

[17] FY 2016 NDAA, Sec. 875.

[18] Van Atta et al., 2017.

[19] There requirements were the 50/50 rule found in 10 U.S.C. 2466 and the "core capabilities rule" in 10 U.S.C. 2464.

## Table A.2. Section 875 Report Recommendations

| Recommendation Number | Summarized Recommendation |
|---|---|
| Recommendation 1 | Make sustainment- and acquisition-related IP data and rights an explicitly stated high priority in DoD management and oversight of acquisition programs. |
| Recommendation 2 | Establish or expand existing organizational capabilities within the DOD components (with OSD support) to provide expertise in the acquisition of IP data and rights to program managers throughout their programs' life cycles and to other staff involved in weapon systems acquisition. |
| Recommendation 3 | Require DOD acquisition programs that are largely dependent on sole-source contracts to OEMs for sustainment to conduct a business case analysis of options to transition to a competitive model for sustainment (maintenance and supply). The results should be forwarded to the component acquisition executive with a recommended plan to obtain the needed IP data and rights. |
| Recommendation 4 | State as a matter of policy that DOD acquisition programs that use commercial derivative aircraft should maximize use of data provided for FAA certified aircraft under FAA regulations to facilitate competition for maintenance and supply of parts for systems and components. |
| Recommendation 5 | Establish under OSD auspices an ongoing DOD advisory group to identify and, in consultation with industry, seek resolution of ambiguities and disagreements in terms and provisions related to DOD sustainment needs, particularly those involving access to and use of IP. The group should be tasked to develop an appendix to DFARS that would specify in greater detail the meaning of such terms as OMIT data, FFF data, and detailed manufacturing and process data. |
| Recommendation 6 | DOD should support and fund an assessment of DOD acquisition and sustainment specifically focused on alternative approaches for contracting and overseeing the development, procurement, and sustainment of weapon systems under severely limited competition. |

SOURCE: Van Atta et al., 2017.

# References

Advisory Panel on Streamlining and Codifying Acquisition Regulations (Section 809 Panel), *Interim Report*, Washington, D.C.: U.S. Department of Defense, May 2017; *Volume 1*, January 2018; *Volume 2*, June 2018; *Volume 3*, January 2019.

Air Force Instruction 63-101/20-101, *Integrated Life Cycle Management*, Washington, D.C.: U.S. Department of the Air Force, May 9, 2017. As of November 27, 2019: https://static.e-publishing.af.mil/production/1/saf_aq/publication/afi63-101_20-101/afi63 -101_20-101.pdf

Anson, Weston, and Donna Suchy, eds., *Intellectual Property Valuation: A Primer for Identifying and Determining Value*, Chicago, Ill.: American Bar Association Section of Intellectual Property Law, 2005.

Arena, Mark V., John Birkler, Irv Blickstein, Charles Nemfakos, Abby Doll, Jeffrey A. Drezner, Gordon T. Lee, Megan McKernan, Brian McInnis, Carter C. Price, Jerry M. Sollinger, and Erin York, *Management Perspectives Pertaining to Root Cause Analyses of Nunn-McCurdy Breaches: Contractor Motivations and Anticipating Breaches*, Vol. 6, Santa Monica, Calif.: RAND Corporation, MG-1171/6, 2014. As of January 27, 2019: https://www.rand.org/pubs/monographs/MG1171z6.html

Ausink, John A., Lisa M. Harrington, Laura Werber, William A. Williams, John E. Boon Jr., and Michael H. Powell, *Air Force Management of the Defense Acquisition Workforce Development Fund: Opportunities for Improvement*, Santa Monica, Calif.: RAND Corporation, RR-1486-AF, 2016. As of February 14, 2019: https://www.rand.org/pubs/research_reports/RR1486.html

Baldwin, Laura H., Frank Camm, Edward G. Keating, and Ellen M. Pint, *Incentives to Undertake Sourcing Studies in the Air Force*, Santa Monica, Calif.: RAND Corporation, DB-240-AF, 1998. As of November 30, 2019: https://www.rand.org/pubs/documented_briefings/DB240.html

Becker, Gina, *Human Systems Integration Competency Development for Navy Systems Commands*, Monterey, Calif.: Naval Postgraduate School, September 2012.

Boito, Michael, Kevin Brancato, John C. Graser, and Cynthia R. Cook, *The Air Force's Experience with Should-Cost Reviews and Options for Enhancing Its Capability to Conduct Them*, Santa Monica, Calif.: RAND Corporation, TR-1184, 2012. As of November 30, 2019: https://www.rand.org/pubs/technical_reports/TR1184.html

Boito, Michael, Tim Conley, Joslyn Fleming, Alyssa Ramos, and Katherine Anania, *Expanding Operating and Support Cost Analysis for Major Programs During the DoD Acquisition Process: Legal Requirements, Current Practices, and Recommendations*, Santa Monica, Calif.: RAND Corporation, RR-2527-OSD, 2018. As of November 30, 2019:
https://www.rand.org/pubs/research_reports/RR2527.html

Boito, Michael, Thomas Light, and Lane F. Burgette, *Relationships Between Aircraft Age and Reliability, Maintenance, and Readiness Outcomes*, Santa Monica, Calif.: RAND Corporation, 2014, Not available to the general public.

Boito, Michael, Thomas Light, Patrick Mills, and Laura H. Baldwin, *Managing U.S. Air Force Aircraft Operating and Support Costs: Insights from Recent RAND Analysis and Opportunities for the Future*, Santa Monica, Calif.: RAND Corporation, RR-1077-AF, 2016. As of May 2, 2020:
https://www.rand.org/pubs/research_reports/RR1077.html

Borowski, Samuel M., *Advance Agreements for Sustainment (FOUO)*, Bullet background paper, Washington, D.C.: Office of the Air Force Deputy General Counsel for Acquisition, September 7, 2018.

Camm, Frank, "Adapting Best Commercial Practices to Defense," in Stuart Johnson, Martin C. Libicki, and Gregory F. Treverton, eds., *New Challenges, New Tools for Defense Decisionmaking*, Santa Monica, Calif.: RAND Corporation, MR-1576-RC, 2003, pp. 211–246. As of November 30, 2019:
https://www.rand.org/pubs/monograph_reports/MR1576.html

Camm, Frank, Jeffrey A. Drezner, Beth E. Lachman, and Susan A. Resetar, *Implementing Proactive Environmental Management: Lessons Learned from Best Commercial Practice*, Santa Monica, Calif.: RAND Corporation, MR-1371-OSD, 2001. As of November 30, 2019:
https://www.rand.org/pubs/monograph_reports/MR1371.html

Camm, Frank, Laura Werber, Julie Kim, Elizabeth Wilke, and Rena Rudavsky, "Implementation of Significant Change in the Inspection System," in *Charting the Course for a New Air Force Inspection System*, Santa Monica, Calif.: RAND Corporation, TR-1291-AF, 2013, pp. 87–100. As of November 30, 2019:
https://www.rand.org/pubs/technical_reports/TR1291.html

Camm, Frank, Thomas C. Whitmore, Guy Weichenberg, Sheng Tao Li, Phillip Carter, Brian Dougherty, Kevin Nalette, Angelena Bohman, and Melissa Shostak, *Data Rights Relevant to Weapon Systems in Air Force Special Operations Command*, Santa Monica, Calif.: RAND Corporation, RR-4298-AF, 2021.

Chenoweth, Mary E., Michael Boito, Shawn McKay, and Rianne Laureijs, *Applying Best Practices to Military Commercial-Derivative Aircraft Engine Sustainment: Assessment of Using Parts Manufacturer Approval (PMA) Parts and Designated Engineering Representative (DER) Repairs*, Santa Monica, Calif.: RAND Corporation, RR-1020/1-OSD, 2016. As of November 30, 2019:
https://www.rand.org/pubs/research_reports/RR1020z1.html

Commission on Army Acquisition and Program Management in Expeditionary Operations, *Urgent Reform Required: Army Expeditionary Contracting*, Washington, D.C.: Secretary of the Army, 2007. As of February 27, 2019:
https://www.acq.osd.mil/dpap/contingency/reports/docs/gansler_commission_report_final_report_20071031.pdf

Cook, Cynthia R., Caroline Baxter, Laura Werber Castaneda, and Abigail Haddad, "Implementation," in Bernard D. Rostker et al., *Sexual Orientation and U.S. Military Personnel Policy: An Update of RAND's 1993 Study*, Santa Monica, Calif.: RAND Corporation, MG-1056-OSD, 2010, pp. 371–388. As of November 30, 2019:
https://www.rand.org/pubs/monographs/MG1056.html

DCMA Instruction 3101, *Program Support*, Washington, D.C.: U.S. Department of Defense, July 28, 2017, change 1, September 20, 2018.

DCMA Manual 3101-01, *Program Support Life Cycle*, Washington, D.C.: U.S. Department of Defense, October 23, 2017, change 1, September 20, 2018.

Defense Contract Audit Agency, "About DCAA," webpage, undated-a. As of January 18, 2019:
https://www.dcaa.mil/Home/AboutDCAA?title=Mission

———, "Qualifications," webpage, undated-b. As of January 18, 2019:
https://www.dcaa.mil/Career/Qualifications

———, *Fiscal Year (FY) 2019 President's Budget: Operation and Maintenance, Defense-Wide*, Fort Belvoir, Va., February 2018a.

———, *Report to Congress on FY 2017 Activities*, Washington, D.C., 2018b.

*Delco Electronics Corporation*, Comptroller General of the United States, B-244559, October 29, 1991, 91-2 CPD ¶ 391.

Department of Defense Instruction 5000.02, *Operation of the Defense Acquisition System*, Washington, D.C.: U.S. Department of Defense, Office of the Under Secretary for Acquisition, Technology, and Logistics, January 7, 2015, p. 6.

DeVecchio, W. Jay, "Data Rights Assault: What in the H (Clause) Is Going on Here? Air Force Overreaching on OMIT Data," *Government Contractor*, Vol. 60, No. 2, January 17, 2018, pp. 1–6.

Director, Acquisition Workforce Management, "ASN(RDA)PCD 1102s Rotations at Price Fighters," Assistant Secretary of the Navy for Research, Development, and Acquisition, Washington, D.C., undated. As of May 1, 2020: https://www.secnav.navy.mil/rda/workforce/Documents/ASN(RDA)%20PCD%201102s %20rotations%20at%20Price%20Fighters%20-%203.docx

Donovan, Matthew P., "Intellectual Property Rights Cross-Functional Team (IPR CFT)," memorandum, Washington, D.C.: Office of the Under Secretary of the Air Force, February 21, 2018.

European IPR Helpdesk, "Fact Sheet: Intellectual Property Valuation," June 2015. As of May 2, 2020: https://www.iprhelpdesk.eu/sites/default/files/newsdocuments/Fact-Sheet-IP-Valuation.pdf

Federal Acquisition Regulation 17.103. Title 48. Federal Acquisition Regulations System. Chapter 1. Federal Acquisition Regulation. Subchapter C. Contracting Methods and Contracting Types. Part 17. Special Contracting Methods. Subpart 17.1. Multiyear Contracting. Section 17.103. Definitions. 48 CFR § 17.103 (2002).

Federal Acquisition Regulation § 31.109, Title 48. Federal Acquisition Regulations System. Chapter 1. Federal Acquisition Regulation. Subchapter E. General Contracting Requirements. Part 31. Contract Cost Principles and Procedures. Subpart 31.1. Applicability. § 31.109, "Advance Agreements," 48 CFR § 31.109 (2014).

*Freightliner Corporation*, Armed Services Board of Contract Appeals (ASBCA), 94-1 BCA 26,538, 1993 WL 502202, ASBCA No. 42,982 (November 26, 1993).

Gallo, Marcy E., "Federally Funded Research and Development Centers (FFRDCs): Background and Issues for Congress," Washington, D.C.: Congressional Research Service, R44629, December 1, 2017. As of January 11, 2019: https://fas.org/sgp/crs/misc/R44629.pdf

Gates, Susan M., Edward G. Keating, Adria D. Jewell, Lindsay Daugherty, Bryan Tysinger, Albert A. Robbert, and Ralph Masi, *The Defense Acquisition Workforce: An Analysis of Personnel Trends Relevant to Policy, 1993–2006*, Santa Monica, Calif.: RAND Corporation, TR-572-OSD, 2008. As of November 30, 2019: https://www.rand.org/pubs/technical_reports/TR572.html

Gates, Susan M., Edward G. Keating, Bryan Tysinger, Adria D. Jewell, Lindsay Daugherty, and Ralph Masi, *The Department of the Navy's Civilian Acquisition Workforce: An Analysis of Recent Trends*, Santa Monica, Calif.: RAND Corporation, TR-555-NAVY, 2009. As of November 30, 2019: https://www.rand.org/pubs/technical_reports/TR555.html

Gates, Susan M., Brian Phillips, Michael H. Powell, Elizabeth Roth, and Joyce S. Marks, *Analyses of the Department of Defense Acquisition Workforce: Update to Methods and Results Through FY 2017*, Santa Monica, Calif.: RAND Corporation, RR-2492-OSD, 2018. As of February 14, 2019:
https://www.rand.org/pubs/research_reports/RR2492.html

Gates, Susan M., Elizabeth Roth, Sinduja Srinivasan, and Lindsay Daugherty, *Analyses of the Department of Defense Acquisition Workforce: Update to Methods and Results through FY 2011*, Santa Monica, Calif.: RAND Corporation, RR-110-OSD, 2013. As of November 30, 2019:
https://www.rand.org/pubs/research_reports/RR110.html

Guo, Christopher, Philip Hall-Partyka, and Susan M. Gates, *Retention and Promotion of High-Quality Civil Service Workers in the Department of Defense Acquisition Workforce*, Santa Monica, Calif.: RAND Corporation, RR-748-OSD, 2014. As of November 30, 2019:
https://www.rand.org/pubs/research_reports/RR748.html

Heritage Foundation, *An Assessment of U.S. Military Power: U.S. Air Force*, Washington, D.C.: Heritage Foundation, October 4, 2018. As of February 14, 2018:
https://www.heritage.org/military-strength/assessment-us-military-power/us-air-force

Insinna, Valerie, "Lockheed Loses Out on Its US Air Force Huey Replacement Protest," *Defense News*, May 22, 2018. As of November 26, 2019:
https://www.defensenews.com/air/2018/05/22/lockheed-loses-out-on-huey-replacement
-protest/

International Organization for Standardization, *Brand Valuation—Requirements for Monetary Brand Valuation, ISO 10668:2010*, Geneva, Switzerland: International Organization for Standardization, September 2010.

Kennedy, Michael, Thomas Light, Michael Boito, Fred Timson, and Haralambos Theologis, *USAF Aircraft Operating and Support Cost Growth*, Santa Monica, Calif.: RAND Corporation, RR-813-AF, 2014.

Kotter, John P., *Leading Change*, Boston: Harvard Business School Press, 1996.

Light, Thomas, Michael Boito, Timothy Conley, Larry Klapper, and John Wallace, *Understanding Changes in U.S. Air Force Aircraft Depot-Level Reparable Costs over Time*, Santa Monica, Calif.: RAND Corporation, RR-2518-AF, 2018.

Light, Thomas, Dwayne M. Butler, Michael Boito, Vikram Kilambi, Kristin J. Leuschner, Sheng Tao Li, Abby Schendt, and Sunny D. Bhatt, *Management of U.S. Air Force Aircraft Contractor Logistics Support Arrangements: Summary of Findings and Recommendations*, Santa Monica, Calif.: RAND Corporation, RR-4194-AF, forthcoming.

Lorell, Mark A., Robert S. Leonard, and Abby Doll, *Extreme Cost Growth: Themes from Six U.S. Air Force Major Defense Acquisition Programs*, Santa Monica, Calif.: RAND Corporation, RR-630-AF, 2015. As of November 30, 2019:
https://www.rand.org/pubs/research_reports/RR630.html

Martin, Bradley, Michael E. McMahon, James G. Kallimani, and Tim Conley, *Accounting for Growth in the Ship Depot Maintenance Account*, Santa Monica, Calif.: RAND Corporation, RR-1837-NAVY, 2017.

Martinez, Luis, and Elizabeth McLaughlin, "How the Pentagon Has Saved $4.7 Billion in the Last Two Years," *ABC News*, January 26, 2019. As of January 29, 2019:
https://abcnews.go.com/beta-story-container/Politics/pentagon-saved-47-billion-past-years/story?id=60623672

McCarthy, Alice, "Transforming the U.S. Naval Systems Command with Thanks to MIT," *MIT News*, August 6, 2018. As of January 17, 2019:
http://news.mit.edu/2018/mit-online-course-model-based-design-transforming-us-dod-navair-acquisitions-0806

Milgrom, Paul, and John Roberts, "The Economics of Modern Manufacturing: Technology, Strategy, and Organization," *American Economic Review*, Vol. 80, No. 3, June 1990, pp. 511–528.

Moore, Nancy Young, Laura H. Baldwin, Frank Camm, and Cynthia R. Cook, *Implementing Best Purchasing and Supply Management Practices: Lessons from Innovative Commercial Firms*, Santa Monica, Calif.: RAND Corporation, DB-334-AF, 2002. As of November 30, 2019:
https://www.rand.org/pubs/documented_briefings/DB334.html

Naval Air Systems Command, "Overview," Patuxent, Md.: U.S. Navy, undated. As of January 18, 2019:
https://web.archive.org/web/20171224130745/http://www.navair.navy.mil:80/index.cfm?fuseaction=home.display&key=9E99EE24-2F3D-4E23-A0C1-A54C18C3FFC8

———, *NAVAIR Acquisition Guide 2014/2015*, Patuxent, Md.: U.S. Navy, October 1, 2013.

———, "1.5 Role of the Competency Aligned Organization," in *NAVAIR—Integrated Program Team Manual: Guidance for Program Teams and Their Subsets December 1996*, Patuxent, Md.: U.S. Navy, April 10, 2014. As of February 13, 2019:
https://web.archive.org/web/20170812144311/http://www.navair.navy.mil/nawctsd/Resources/Library/Acqguide/navipt1.htm

Naval Air Warfare Center, "Funding Documents," Orlando, Fla.: U.S. Navy, Training System Division, August 15, 2013. As of January 21, 2019:
https://web.archive.org/web/20170929170627/http://www.navair.navy.mil/nawctsd/Resources/Library/Acqguide/funddocs.htm

Naval Supply Systems Command Weapon Systems Support, *Navy Price Fighters*, Norfolk, Va.: Naval Supply Systems Command Weapon Systems Support, February 2018.

Office of Management and Budget, *Guidelines and Discount Rates for Benefit-Cost Analysis of Federal Programs*, Circular A-94, Transmittal Memo No. 64, Washington, D.C.: Office of the President, undated. As of November 26, 2019:
https://www.whitehouse.gov/sites/whitehouse.gov/files/omb/circulars/A94/a094.pdf

———, *Discount Rates for Cost-Effectiveness, Lease Purchase, and Related Analyses*, Circular No. A-94 Appendix C, Washington, D.C.: Office of the President, November 2017. As of January 28, 2019:
https://www.whitehouse.gov/wp-content/uploads/2017/11/Appendix-C.pdf

Office of the Deputy Assistant Secretary for Defense for Systems Engineering, *Reliability, Availability, Maintainability, and Cost (RAM-C) Rationale Report Outline Guidance*, Version 1.0, Washington, D.C.: Office of the Secretary of Defense, February 28, 2017. As of February 4, 2019:
https://ac.cto.mil/wp-content/uploads/2019/06/RAMC-Outline-2017-07-07.pdf

Office of the Joint Staff, J-8, *Manual for the Operation of the Joint Capabilities Integration and Development System*, Washington, D.C.: Department of Defense, August 31, 2018. As of February 4, 2019:
https://www.dau.mil/cop/rqmt/DAU%20Sponsored%20Documents/Manual%20-%20JCIDS,%2031%20Aug%202018.pdf

President's Blue Ribbon Commission on Defense Management, *A Quest for Excellence: Final Report to the President*, Washington, D.C., 1986. As of February 27, 2019:
https://assets.documentcloud.org/documents/2695411/Packard-Commission.pdf

Schwartz, Moshe, and Charles V. O'Connor, *The Nunn-McCurdy Act: Background, Analysis, and Issues for Congress*, Washington, D.C.: Congressional Research Service, R41293, May 12, 2016.

Section 809 Panel Report—*see* Advisory Panel on Streamlining and Codifying Acquisition Regulations (Section 809 Panel).

Section 813 Panel Report—*see* U.S. Department of Defense, *2018 Report of the Government-Industry Advisory Panel on Technical Data Rights*.

Smith, Len, *Valuation of Intellectual Property*, ISYM 540, Current Topics in ISM, July 2, 2009.

Springsteen, Beth, and Elizabeth K. Bailey, *The F/A-18E/F: An Integrated Product Team (IPT) Case Study*, Alexandria, Va.: Institute for Defense Analysis, NS D-8027, April 1998.

Stecher, Brian, Frank Camm, Cheryl Damberg, Laura Hamilton, Kathleen Mullen, Christopher Nelson, Paul Sorenson, Martin Wachs, Allison Yoh, and Gail Zellman, *Toward a Culture of Consequences: Performance-Based Accountability Systems for Public Services*, Santa Monica, Calif.: RAND Corporation, MG-1019, 2010. As of May 2, 2020: https://www.rand.org/pubs/monographs/MG1019.html

U.S. Air Force, Space and Missile Systems Center, *Acquiring and Enforcing the Government's Rights in Technical Data and Computer Software under Department of Defense Contracts: A Practical Handbook for Acquisition Professionals*, 9th ed., Los Angeles Air Force Base, Ca.: U.S. Air Force, October 2018.

U.S. Department of Defense, *2018 Report of the Government-Industry Advisory Panel on Technical Data Rights*, Washington, D.C.: U.S. Department of Defense, November 13, 2018.

U.S. General Accounting Office, *Defense Contract Audits: Defense Contract Audit Agency's Staff Qualifications, Experience, Turnover, and Training*, B-241812, Washington, D.C.: U.S. General Accounting Office, July 1991.

———, *Federally Funded R&D Centers: Information on the Size and Scope of DOD-Sponsored Centers*, Washington, D.C.: U.S. General Accounting Office, National Security and International Affairs Division, B-270464, April 24, 1996. As of January 11, 2019: https://www.gao.gov/assets/230/222626.pdf

U.S. House of Representatives, *National Defense Authorization Act for Fiscal Year 2018 Conference Report to Accompany H.R. 2810*, Washington, D.C., November 2017.

Van Atta, Richard, Royce Kneece, Michael Lippitz, and Christina Patterson, *Department of Defense Access to Intellectual Property for Weapon Systems Sustainment*, IDA Paper P-8266, Alexandria, Va.: Institute for Defense Analyses, May 2017.

White, James W., "Application of New Management Concepts to the Development of F/A-18 Aircraft," *Johns Hopkins APL Technical Digest*, Vol. 18, No. 1, 1997, pp. 21–32.

Wirtz, Harald, "Valuation of Intellectual Property: A Review of Approaches and Methods," *International Journal of Business and Management*, Vol. 7, No. 9, May 2012, pp. 40–48.

Younossi, Obaid, David E. Stem, Mark A. Lorell, and Frances M. Lussier, *Lessons Learned from the F/A-22 and F/A-18E/F Development Programs*, Santa Monica, Calif.: RAND Corporation, MG-276, 2005. As of November 30, 2019: https://www.rand.org/pubs/monographs/MG276.html

Zellman, Gail L., Joanna Zorn Heilbrunn, Conrad Schmidt, and Carl H. Builder, "Implementing Policy Change in Large Organizations," in Bernard D. Rostker et al., *Sexual Orientation and U.S. Military Personnel Policy: Options and Assessment*, Santa Monica, Calif.: RAND Corporation, MR-323-OSD, 1993, pp. 368–394. As of November 30, 2019: https://www.rand.org/pubs/monograph_reports/MR323.html